THE BUREAUCRACY
OF EMPATHY

CORPUS

The Humanities in Politics and Law

JURIS

Series editor: Elizabeth S. Anker, Cornell University

CORPUS JURIS: THE HUMANITIES IN POLITICS AND LAW PUBLISHES BOOKS AT THE INTERSECTIONS BETWEEN LAW, POLITICS, AND THE HUMANITIES—INCLUDING HISTORY, LITERARY CRITICISM, ANTHROPOLOGY, PHILOSOPHY, RELIGIOUS STUDIES, AND POLITICAL THEORY. BOOKS IN THIS SERIES TACKLE NEW OR UNDER-ANALYZED ISSUES IN POLITICS AND LAW AND DEVELOP INNOVATIVE METHODS TO UNDERTAKE THOSE INQUIRIES. THE GOAL OF THE SERIES IS TO MULTIPLY THE INTERDISCIPLINARY JUNCTURES AND CONVERSATIONS THAT SHAPE THE STUDY OF LAW.

THE BUREAUCRACY OF EMPATHY

Law, Vivisection, and Animal Pain in Late Nineteenth-Century Britain

Shira Shmuely

CORNELL UNIVERSITY PRESS   ITHACA AND LONDON

First published 2023 by Cornell University Press

Library of Congress Cataloging-in-Publication Data

Names: Shmuely, Shira, 1982– author.
Title: The bureaucracy of empathy : law, vivisection, and animal pain in late nineteenth-century Britain / Shira Shmuely.
Description: Ithaca [New York] : Cornell University Press, 2023. | Series: Corpus juris: the humanities in politics and law | Includes bibliographical references and index.
Identifiers: LCCN 2023000918 (print) | LCCN 2023000919 (ebook) | ISBN 9781501770388 (hardcover) | ISBN 9781501770395 (paperback) | ISBN 9781501770401 (epub) | ISBN 9781501770418 (pdf)
Subjects: LCSH: Great Britain. Cruelty to Animals Act of 1876. | Animal welfare—Great Britain—History—19th century. | Animal welfare—Law and legislation—Great Britain.
Classification: LCC HV4805.A3 S56 2023 (print) | LCC HV4805.A3 (ebook) | DDC 179/.30941—dc23/eng/20230324
LC record available at https://lccn.loc.gov/2023000918
LC ebook record available at https://lccn.loc.gov/2023000919

To my mother, Dalia Shmuely,
and in loving memory of my father,
Mordechai (Moti) Shmuely

CONTENTS

ACKNOWLEDGMENTS

I have always had a raw and intuitive interest in animals and the ways people around me treat them. I was perplexed by how moments of strong affection could turn into cold disregard, but I could not make sense of it. Shai Lavi's seminar at Tel Aviv University about animals in law and jurisprudence—a real rarity in the 2000s—drew me into the then extremely small niche in the humanities in which I could explore these relationships. While enrolled in MIT's doctoral program in History, Anthropology and Science, Technology, and Society (HASTS), I was lucky to learn from Harriet Ritvo, one of the pioneers who made the niche into a thriving field. My greatest debt is to her dedicated mentorship. From beginning to end, a word would not have been written without her. Stefan Helmreich deserves ocean-deep gratitude for encouraging me to dare while always being "in conversation with." Christopher Capozzola guided me to "let the actors talk," and his intellectual enthusiasm revived me more than once.

Generous funds for this project were provided by the National Science Foundation, the Council for European Studies, the Center for International Studies at MIT, and the Kelly Douglas Fund at MIT. During the research, I came to admire the work of many librarians and archivists who have mastered and preserved the foundations of historical knowledge. Most of the collections I consulted are held by the British National Archives and the Wellcome Library, London. Other valuable materials are at the Royal College of Physicians, London, the library of the Royal College of Surgeons of England, the British Library, and the Countway Library of Medicine at Harvard University. Specifically, I am thankful to Michelle Baldwin from MIT libraries, who guided my first steps in this project.

I am thankful for the support of the History Department at MIT, especially Mabel Chin Sorett, Margo Collett, Lerna Ekmekçioğlu, Jeffery Ravel, Craig Wilder, and Elizabeth Wood. The HASTS program at MIT was my home even when I was away, and the stewardship of Karen Gardner was invaluable in all aspects. I deeply thank my teachers Jose Brunner, Natasha Dow Schull, Anita Guerrini, Sheila Jasanoff, David Jones, David Kaiser, Clapperton Mavhunga, Heather Paxson, Hanna Rose Shell, Susan Silbey, and Chris Walley. As a visiting researcher at the Georgetown University Law Center I benefited from the brilliant minds of Robin West and Gregory Class, who reconnected me with issues of the present.

Splendid friends provided their bright ideas to earlier drafts. They all made the craft joyful and memorable: Renee Blackburn, Marie Burks, Amah Edoh, Shreeharsh Kelkar, Nicole Labruto, Moran Levy, Lan Li, Daniel Mann, Liron Mor, Lucas Mueller, Oded Na'aman, Tamar Novik, Oded Rabinovitch, Joanna Radin, Caterina Scaramelli, Alma Steingart, Michelle Spektor, Maayan Sudai, Mitali Thakor, and Lihi Yona. I gained a lot from the genuine generosity of Canay Özden-Schilling, Tom Özden-Schilling, and Rebecca Woods.

During the years of working on this project, I touched life from both ends. I lost my father and gave birth to my children. This book is dedicated to my dad, who dreamt that I would be a scientist, and compromised with me writing about science, and to my admirable mother, who traveled the world to support my work. To my kids, Nitai, Mori, and Rafael—thank you for bringing the old awe and wonder of the animal kingdom back into my life. The Mann family was a significant anchor in the transatlantic water. None of my siblings—Naama, Hemdat, and Nadav—read a word that I wrote, yet their tribal love and cheerful humor kept me going. And above all, thank you, Itamar, a true believer.

THE BUREAUCRACY OF EMPATHY

INTRODUCTION

A well-dressed gentleman stands with his back to us, facing a small black dog, perhaps a terrier. The fingers of the man's left hand are spread out as he leans on the desk on which the dog sits. His right hand is behind his back, gripping a glass bottle. He is positioned so that only we, the viewers, can see the small container. Next to the dog lies an assortment of surgical instruments, and behind him is a microscope. The dog cannot see what the man has behind his back, but his expression betrays anxiety as he puts his paws together in submission, almost begging. His fate is as clear as the liquid in the bottle. The caption of the engraving in its 1885 London exhibition was "Vivisection—The Last Plea" (Figure 1). The engraving visualizes the practice of animal experimentation following the British Cruelty to Animals Act (1876). Among other provisions, the Act, also known as the Vivisection Act, outlawed public demonstrations of experiments and obliged experimenters to hold a license and practice in registered places.[1] The new order set by the Act, which required the use of anesthesia, is represented by the small bottle held by the experimenter in the engraving. Under the authority of the Home Office, the 1876 Act was in force for more than a century and inspired similar legislation in other countries.

The Vivisection Act created a comprehensive administrative system to govern, in the words of Section 2, "any experiment calculated to give pain to a living animal." No pain—no law. After 1876, if a British physiologist planned an experiment that would cause pain to the animals being used, a whole set of guidelines came into play. The physiologist, almost always a man in that period, would have to submit a license application to the Home Office. Depending on the intended procedure, he might also have to submit a special certificate signed by two leading medical figures. The law-abiding physiologist would register his laboratory at the Home Office, thereby putting an end to weekend experiments in his home basement. While running the experiment, he would be required, among other things, to use anesthesia and kill the animals before their recovery from its effects (a Home Office inspector might walk into the laboratory without prior notice).

---

1. Cruelty to Animals Act, 1876, 39 & 40 Vict. c. 77.

FIGURE 1 / "A Dog on a Laboratory Bench Sits Up and Begs the Prospective Vivisector for Mercy." Engraving by C. J. Tomkins, 1883, after a painting by J. McClure Hamilton. Wellcome Collection, 25933i.

Upon concluding the experiment, the physiologist's assistants would record the details of the procedure in anticipation of a future inspection by the Home Office.

However, the same post-1876 British physiologist did not need to notify the Home Office if his planned experiments were regarded painless. He could then go ahead with his research routine in his university laboratory or home basement, undisturbed by the Act and its administration in all its tedious forms. The definition of pain was, therefore, crucial for determining one's research strategies and their legality, as much as it was essential for designating the responsibilities of the Home Office under the Vivisection Act.

But what was pain? Having a certain idea of pain, a way to detect it, measure it, and talk about it, were fundamental for anyone engaged with the Vivisection Act, be the person a scientist, a civil servant, or an animal lover watching over the implementation of the Act. Nevertheless, the Act did not define "pain," just as it did not define "experiment," and left it for bureaucrats, scientists, antivivisectionists, and jurists to debate. Their exchanges, negotiations, and formulations of pain make up the case studies in this book. The period under inquiry starts with the 1875 Royal Commission on Vivisection, which recommended the enactment of the law. The narrative ends three decades later, with the report of the second Royal Commission on Vivisection (1906–12), which concluded three formative decades of developing the implementation policy of the Home Office.

In discussing the significance of the Vivisection Act, I make three core arguments. First, the Act molded ideas of moral conduct in animal experimentation into a set of legal provisions, which infiltrated almost every aspect of the physiology profession and shaped experimental medicine for a century. When civil servants guided experimenters in modifying their practices in accordance with the regulations for the prevention of cruelty to animals, they intertwined practical and normative aspects of animal research. Late nineteenth-century law was inseparable from physiology, and normative order played a crucial role in the production of scientific facts in biomedical research. The Act became one of the factors that defined and shaped experiments. If experimentation produced scientific facts, regulated animal experimentation produced scientific facts with the additional administrative assurance that the knowledge embodied in the published facts had been obtained by morally acceptable means. Seeing the extent of state intervention in physiologists' methods, be it through the requirement to fill out forms or the obligation to put down a suffering animal, is essential not only for our understanding of the broad reach of law but also for a clearer view of the structure of modern biomedical science.[2]

2. The result of the intervention of the Home Office in experimental practice was the introduction of an ostensibly cruelty-free fact, or ethical scientific fact. See Shira Shmuely, "Law and the Laboratory: The British Vivisection Inspectorate in the 1890s," *Law & Social Inquiry* (April 8, 2021), 1–31.

Second, as much as the history of animal experimentation regulation is that of the incorporation of legal norms into the life sciences, it is also a story about science affecting legal concepts and practices. Scientific claims about physiological phenomena directed British regulators when drafting the Act, and informed civil servants when creating policies and conducting routine enforcement. Late nineteenth-century scientific ideas of pain, although in constant flux, framed the ways legislators, civil servants, and lawyers understood sentience and responded to sentient animal bodies. Anesthesia, for example, was adopted as a dominant tool of the law. Furthermore, while legal representatives pronounced their ideas about animal pain, men of science provided their legal interpretation of the Act. Scientific theories and ethical norms were circulating between the worlds of law and science.

Third, the Act imposed a utilitarian relationship with experimental animals. While it traded nonhuman animals' agonies for scientific progress, it also mandated a relation of care and attention to the pain of these creatures. Moreover, the definition of pain as the prism through which the use of animals was evaluated generated a process of empathy. Contrary to the prevailing view in the history of laboratories and animal studies, animals in regulated research environments did not become yet another item in a set of artifacts.[3] With the Vivisection Act in force, almost any engagement with research animals led those involved to evaluate the sensory experiences of the animal subjects through a variety of cognitive and emotional processes. The Home Office administration attempted to mold these processes into bureaucratic forms and generalized guidelines.

To elucidate the dynamic between law and science as it played out in the regulation of vivisection, I introduce two theoretical concepts: bureaucracy of empathy and the coproduction of science and normative order.

The Act established a bureaucracy of empathy. Bureaucracy here stands for its plainest definition as "a system of government or . . . administration by a hierarchy of professional administrators following clearly defined procedures in a routine and organized manner."[4] Home Office bureaucracy encompasses

---

3. Robert E. Kohler, *Lords of the Fly: Drosophila Genetics and the Experimental Life* (Chicago: University of Chicago Press, 1994); Hans-Jörg Rheinberger, *An Epistemology of the Concrete: Twentieth-Century Histories of Life* (Durham: Duke University Press, 2010); Karen Rader, *Making Mice: Standardizing Animals for American Biomedical Research, 1900–1955* (Princeton, NJ: Princeton University Press, 2004); Rachel A. Ankeny and Sabina Leonelli, "What's so Special about Model Organisms?," *Studies in History and Philosophy of Science* 42, no. 2 (June 2011), 313–23.

4. *Oxford English Dictionary Online*, s.v. "Bureaucracy," http://www.oed.com/view/Entry/24905?redirectedFrom=bureaucracy. Max Weber famously identified bureaucracy as the distinctive mark of modernity. Max Weber, *Economy and Society*, ed. Guenther Roth and Claus Wittich (Berkeley: University of California Press, 2013), ii, 956–58. For the way filling out forms produces citizenship, see

the procedures, routines, and ideas that were instituted following legislation to implement the overarching principle of managing the pain of animals in laboratories. Empathy stands for feeling what one imagines the other feels or should feel. In the context of the Act, it is the scientist's or bureaucrat's ability to identify with nonhuman animals and to evaluate their sensory experiences. The actors at the time, it should be noted, did not use the term.

The legal scholar Robin West draws from Adam Smith when she explains, "Empathy tells us, perhaps, something about what others are feeling, or at least gives us a hint of its feel. It is a source of information." Empathy allows the empathizer "access to a certain kind of knowledge—knowledge of the perspective of others."[5] I do not contrast empathy with anatomical and physiological knowledge; to the contrary, the latter was at times integral to the empathizing process. Science was not an objective alternative to empathy, and knowledge sometimes informed attempts to empathize with the animal. It is important to note that this definition of empathy is devoid of moral overtones. Empathy, suggests the philosopher Stephen Darwall, "can be consistent with the indifference of pure observation or even the cruelty of sadism. It all depends on why one is interested in the other's perspective."[6] Empathy can thus be factored into the rational, calculative processing of other creatures' pain with no contradiction in terms. This point becomes clearer when empathy is contrasted with sympathy, which implies a positive and supportive affinity with another being. Again I follow West, who explains that sympathy "is the moral sentiment that aligns our interest with that of another in pain . . . Empathy is not what motivates action. Sympathy is what motivates action." Or, in the words of the historian Rob Boddice, sympathy is "an emotional recognition of the plight of the object in question . . . the actions done in the name of sympathy are, or are supposed to be, good."[7]

The understanding of empathy as a source of knowledge differs from the definition employed by the philosopher and animal studies scholar Lori Gruen, who distinguishes empathy and sympathy in terms of "connection." For Gruen,

---

Christopher Capozzola, *Uncle Sam Wants You: World War I and the Making of the Modern American Citizen* (New York: Oxford University Press, 2010), 23.

5. Robin West, "The Anti-Empathic Turn," in *Passions and Emotions*, ed. James E. Fleming (New York: New York University Press, 2012), 248.

6. Stephen Darwall, "Empathy, Sympathy, Care," *Philosophical Studies: An International Journal for Philosophy in the Analytic Tradition* 89, no. 2–3 (1998): 261. Various, often contradictory, understandings of empathy have developed in diverse intellectual fields. For a survey, see Amy Coplan and Peter Goldie, introduction to *Empathy: Philosophical and Psychological Perspectives*, ed. Amy Coplan and Peter Goldie (Oxford: Oxford University Press, 2014), ix–xlviii.

7. Rob Boddice, *The Science of Sympathy: Morality, Evolution, and Victorian Civilization* (Urbana: University of Illinois Press, 2016), 4.

sympathy "involves maintaining one's own attitudes and adding to them a concern for another." It is "a response to something bad, untoward, unfortunate, or unpleasant happening to someone else in which the individual sympathizing retains all of her attitudes, beliefs, feelings, etc." Empathy, by contrast, "recognizes connection with and understanding of the circumstances of the other," an attunement to the well-being of another.[8] Empathizing with nonhuman animals, according to Gruen, involves immersion in their suffering and perspective. My research has shown that late nineteenth-century British administrators and scientists empathized with animals not to connect but to adjust their use in accordance with the ethical roadmap of the Act.

The second theoretical concept that underlies this book is the coproduction of science and normative order. Starting in the 1980s, historians of science have been preoccupied with investigating the social construction of scientific facts. They have joined other scholars under the aegis of the multidisciplinary field of science and technology studies (or science, technology, and society, both abbreviated STS) to inquire how statements about nature are being made. STS scholars strive to untangle every knot in the knowledge-making process. This literature asks, among other things, what the objects of scientific inquiry are, how research spaces are defined, who is involved in the experimental process and who is left out, what role is played by tools in empirical research, how experiments are designed, and when they end. STS literature also examines the products of scientific inquiries, looking at how knowledge is generalized and universalized, how scientific information comes to be trusted, when it fails to be accepted by specific communities or the broader public, and so forth.[9] The scientific fact was hitherto shown to be an assemblage of elements such as actors, funds, devices, witnessing practices, recording and publication standards, and translation techniques. A close look at the history of vivisection reveals that another substantial

---

8. Lori Gruen, *Entangled Empathy: An Alternative Ethic for Our Relationships with Animals* (New York: Lantern Books, 2015), 44–45.

9. The canonical text in this line of inquiry is Steven Shapin and Simon Schaffer, *Leviathan and the Air-Pump: Hobbes, Boyle, and the Experimental Life* (Princeton, NJ: Princeton University Press, 1985). A noncomprehensive list of major publications in STS and the life sciences that exemplifies these themes includes Harry M. Collins, *Changing Order: Replication and Induction in Scientific Practice* (London: Sage, 1985); Bruno Latour and Steve Woolgar, *Laboratory Life: The Construction of Scientific Facts* (Princeton, NJ: Princeton University Press, 1986); Geoffrey C. Bowker and Susan Leigh Star, *Sorting Things Out: Classification and Its Consequences* (Cambridge: MIT Press, 2000); Soraya de Chadarevian, *Designs for Life: Molecular Biology after World War II* (Cambridge: Cambridge University Press, 2002); Lorraine J. Daston and Peter Galison, *Objectivity* (Princeton, NJ: Zone Books, 2010); Stefan Helmreich, *Alien Ocean: Anthropological Voyages in Microbial Seas* (Berkeley: University of California Press, 2009).

factor in the production of facts in the life sciences, and perhaps also in other scientific disciplines, is normative values or ethics.

STS scholarship shows that scientific apparatuses not only create facts but also organize social relations. These are the workings of coproduction, defined by the STS scholar Sheila Jasanoff as "the simultaneous production of knowledge and public order."[10] Existing STS literature examines public order mainly in the sense of how citizenship is practiced and how state institutions and civil organizations are formed; current literature on coproduction is mostly dedicated to questions of science and democratic politics, and focuses on issues of risk, representation, and allocation of resources.[11] At the same time, the intervention of the state in scientific practice in the name of moral values reveals the normative dimension of law and the ethical aspect of regulation.[12]

Beyond showing that law affected scientific practice and reasoning and that scientists, in turn, influenced administrative and legal decisions regarding the Act, this book describes the ways in which scientists and state bureaucrats were involved in the coproduction of knowledge and normative order. Ensembles of scientists, members of parliament, civil servants, and law officers debated to define the type and intensity of pain that the law focused on, as well as how it was to be recognized. The coproduction by law and science of a definition of pain also involved redefining and redesigning physiological experiments, demonstrating that "the interaction between law and science ends up recreating the world, not only materially but also culturally and morally."[13] To paraphrase the historians Steven Shapin and Simon Schaffer, this book shows that solutions to the problem of knowledge were embedded within practical solutions to the problem

---

10. Sheila Jasanoff, "Beyond Epistemology: Relativism and Engagement in the Politics of Science," *Social Studies of Science* 26, no. 2 (1996): 393.

11. Illuminating examples include Massimiano Bucchi and Federico Neresini, "Science and Public Participation," in *The Handbook of Science and Technology Studies*, edited by Edward J. Hackett et al., 3rd ed (Cambridge: MIT Press, 2007), 453; Kaushik Sunder Rajan, ed., *Lively Capital: Biotechnologies, Ethics, and Governance in Global Markets* (Durham: Duke University Press, 2012); Stephen Hilgartner, Clark Miller, and Rob Hagendijk, eds., *Science and Democracy: Making Knowledge and Making Power in the Biosciences and Beyond* (New York: Routledge, 2015). Even when the sociologist of law Alfons Bora discusses "technoscientific normativity," he centers on the absence of democratic participation in science policy. See Alfons Bora, "Technoscientific Normativity and the 'Iron Cage' of Law," *Science, Technology, & Human Values* 35, no. 1 (2010): 14.

12. In the definition of the Oxford English Dictionary, an ethical act "conforms to moral principles or ethics; morally right; honourable; virtuous; decent; spec. conforming to the ethics of a profession" (*Oxford English Dictionary Online*, s.v. "ethical," https://www.oed.com/view/Entry/64756 ?redirectedFrom=ethical#eid).

13. Susan Silbey and Patricia Ewick, "The Architecture of Authority: The Place of Law in the Space of Science," in *The Place of Law*, ed. Austin Sarat, Lawrence Douglas, and Martha Merrill Umphrey (Ann Arbor: University of Michigan Press, 2003), 78.

of normative order, and practical solutions to the problem of ethics encapsulated practical solutions to the problem of knowledge.[14]

The role of law in this history is not that of an arbitrator between contesting scientific views.[15] Representatives of the law did not have the final say in the disputes surrounding the implementation of the Act. For instance, the magistrate that decided in an 1881 case against a distinguished neurologist, or the General Solicitors' opinions regarding the classification of inoculation procedures, never ended the debates about the meaning of pain or experiment. Legal and research spaces became ad hoc "trading zones," to borrow the historian of science Peter Galison's terminology, where the different cultures of epistemology and reasoning cultivated by law and science came to terms with each other over the questions of pain and experiments.[16] Courtrooms as well as animal shelters at the laboratories were trading zones where physiologists and bureaucrats disputed (and sometimes agreed on) the definitions of pain and experiment while simultaneously informing one another about normative values and knowledge concerning animal bodies. The jointly shaped concepts and practices would dominate the regulation of animal experimentation for a century.

This book historicizes the philosophical conundrums of what is pain and how one knows about others' pain, and examines how British scientists and bureaucrats at the turn of the twentieth century, who were tasked with reflecting upon the inner experiences of nonhuman animals, approached this epistemological, administrative, and physiological challenge. Pain and pleasure were prominent concepts in modern English moral thought. Jeremy Bentham, the emblematic utilitarianist thinker, joined a long tradition originating in Greek philosophy that founded moral principles on pain and pleasure.[17] Educated Victorians were

---

14. Shapin and Schaffer show how "solutions to the problem of knowledge are embedded within practical solutions to the problem of social order, and that different practical solutions to the problem of social order encapsulate contrasting practical solutions to the problem of knowledge" (*Leviathan and the Air-Pump*, 15). For the convergence of "methodologically valid" and "ethically just," see Noortje Jacobs, "A Moral Obligation to Proper Experimentation: Research Ethics as Epistemic Filter in the Aftermath of World War II," *Isis* 111, no. 4 (December 2020): 759–80; Tone Druglitrø, "'Skilled Care' and the Making of Good Science," *Science, Technology, and Human Values* 43, no. 4 (2018): 649–70.

15. The historian of science Graham Burnett evokes such an image of law when examining a whale classification dispute that reached a legal tribunal. D. Graham Burnett, *Trying Leviathan: The Nineteenth-Century New York Court Case That Put the Whale on Trial and Challenged the Order of Nature* (Princeton, NJ: Princeton University Press, 2010).

16. Peter Galison, "Trading Zone: Coordinating Action and Belief," in *The Science Studies Reader*, ed. Mario Biagioli (New York: Routledge, 1999), 137–60, 146.

17. Pain was often discussed in its relation to pleasure and desire, as shown in J. C. B. Gosling and C. C. W. Taylor, *The Greeks on Pleasure* (Oxford: Oxford University Press, 1982); J. C. B. Gosling, *Pleasure and Desire: The Case of Hedonism Reviewed* (Oxford: Oxford University Press, 1969); William

familiar with these ancient texts, as classical training in secondary school was a prerequisite for admission to university.[18] However, the historian of ancient Western philosophy Richard Sorabji notes that in antiquity, the debate about animal ethics centered on the rationality of animals, or lack thereof. Only rarely, and relatively late, did the Greek philosophers mention pain as a reason for treating animals well.[19] Once pain was established as a central element in eighteenth-century moral thought regarding animals, it remained for scholars and scientists to define it, measure it, and fit it into the utilitarian calculus.

Was pain transmissible at all? The literary scholar Elaine Scarry portrays (human) pain as a private experience, before and beyond verbal expression, inherently untransmittable.[20] The historian Joanne Bourke opposes Scarry's interpretation, claiming that pain, as a "type of event," is "infinitely sharable" through imagination and identification.[21] It is at once an existentially absorbing experience, a social matter that generates solidarity, and a racialized and gendered object of governance that is to be located, treated, or manipulated.[22]

Medical organizations today bring into the definition of pain its subjective experience as well as its objective indicators. The latest definition by the International Association for the Study of Pain is: "An unpleasant sensory and emotional experience associated with actual or potential tissue damage, or described in terms of such damage."[23] However, the association complicates this by stating that neither the physiological nor the demonstrative element of the definition is obligatory: "Pain and nociception are different phenomena. Pain

---

V. Harris, *Pain and Pleasure in Classical Times* (Leiden: Brill, 2018). Rob Boddice, "Introduction: Hurt Feelings?," in *Pain and Emotion in Modern History*, ed. Rob Boddice (Houndmills: Palgrave Macmillan, 2014), 6–7.

18. Frank M. Turner, "Victorian Classics: Sustaining the Study of the Ancient World," in *The Organisation of Knowledge in Victorian Britain*, edited by Martin Daunton (Oxford: Oxford University Press, 2005), 160.

19. Richard Sorabji, *Animal Minds and Human Morals: The Origins of the Western Debate* (Ithaca, NY: Cornell University Press, 1995).

20. Elaine Scarry, *The Body in Pain* (Oxford: Oxford University Press, 1988), 4–5.

21. Joanna Bourke, *The Story of Pain: From Prayer to Painkillers* (Oxford: Oxford University Press, 2014), 5, 54.

22. On the political role of pain and the construction of the sufferer, see Adriana Petryna, *Life Exposed: Biological Citizens after Chernobyl* (Princeton, NJ: Princeton University Press, 2013); Keith Wailoo, *Pain: A Political History* (Baltimore: Johns Hopkins University Press, 2016); Martin S. Pernick, *A Calculus of Suffering: Pain, Professionalism and Anesthesia in Nineteenth-Century America* (New York: Columbia University Press, 1987); Lucy Bending, *The Representation of Bodily Pain in Late Nineteenth-Century English Culture* (Oxford: Oxford University Press, 2000).

23. "IASP Terminology," International Association for the Study of Pain, 2020, https://www.iasp-pain.org/Education/Content.aspx?ItemNumber=1698#Pain. See Bourke's discussion of the IAPA's definition in *The Story of Pain*, 12.

cannot be inferred solely from activity in sensory neurons," and "verbal description is only one of several behaviors to express pain; inability to communicate does not negate the possibility that a human or a nonhuman animal experiences pain." Two centuries of systematic scientific study of pain led researchers to incorporate its intractability into its definition: other creatures' pain cannot be inferred only by the activity of their sensory neurons, but it is also not necessarily communicable.[24]

When pain is examined in an administrative context, the aspect of its transmissibility is refined. Victorian vivisection law presumed that animal pain could be traced and identified, and it entrusted civil servants and physiologists with the task of hunting it down. The archives show that the efforts to imagine and understand animal pain provided local and temporary solutions to the problem of pain while repeatedly failing to provide a conclusive medico-legal paradigm for its elimination.

The notion that animals could feel pain was not a nineteenth-century discovery. Indeed, René Descartes famously analogized animals to machines, and their agonized cries to the sound of a broken clock. But the prevalent belief that the Cartesian view represented common seventeenth-century wisdom, and that it was this wisdom that laid the philosophical foundation for animal experimentation, was, according to the historian Anita Guerrini, "simply not true."[25] The majority of the French Academy of Sciences ignored the beast-machine theory, and most anatomists believed that animals could feel pain although they lacked a rational soul.[26] Animal pain had thus been acknowledged before, but what was unique about nineteenth-century Britain was the introduction of new theories about the physiological mechanisms of sensation, and the elevation of pain into a central moral criterion in light of which human–animal relations were to be formed.

Following the publication of Harriet Ritvo's 1987 *Animal Estate*, which treated human–animal relations as a subject worthy of focused historical inquiry, the British context became a favorite choice for investigating the shifting relations

---

24. For theories of pain from diverse perspectives, see Jennifer Corns, ed., *The Routledge Handbook of Philosophy of Pain* (London: Routledge, 2017).

25. Anita Guerrini, *The Courtiers' Anatomists: Animals and Humans in Louis XIV's Paris* (Chicago: University of Chicago Press, 2015), 3; Erica Fudge, *Brutal Reasoning: Animals, Rationality, and Humanity in Early Modern England* (Ithaca, NY: Cornell University Press, 2006), 154–55.

26. In antiquity, the debate about animal ethics centered on the rationality of animals—or the lack thereof. Only rarely, and relatively late, did the Greek philosophers mention pain as a reason for treating animals well. Reason continued to be the distinguishing trait between humans and nonhumans for early modern thinkers. See Sorabji, *Animal Minds and Human Morals*; Fudge, *Brutal Reasoning*.

between humans and other animals.[27] This is, among other reasons, because in British society the move from disinterest in animals' fate to a national concern was radical and evident. The rising sentiments toward animals were influenced by changes in living conditions and were nourished by a complex set of ideas about nature, social order, religion, and nationality. The historian Keith Thomas interprets the increased concern for animals as part of an ongoing reconfiguration of nature in British culture. Thomas attributes the change to the "destruction of the old anthropocentric illusion" with the expansion of knowledge about the world.[28] Another factor was the accelerated urbanization that separated city dwellers from their rural past. Care for animals offered a measure of reconciliation with this rupture.[29]

The growing sentiments toward nonhuman animals in Britain have been well documented by now, and extensive explanations have been offered for this phenomenon. Equally well researched is the massive increase in the use of animals in modernity, especially in the agricultural sector, which responded to a growing demand for meat.[30] Perhaps second to agriculture in animal use, is the medical sector. As the historian Abigail Woods maintains, "Different roles have provided animals with different opportunities to shape medicine, with ramifications for the health of both humans and animals. The concept of the animal role therefore offers a useful tool for illuminating the historical co-constitution of humans, animals and medicine."[31] My contribution to this literature is in exploring the nexus of animal use and animal ethics. I take the Vivisection Act

27. Harriet Ritvo, *The Animal Estate: The English and Other Creatures in the Victorian Age* (Cambridge, MA: Harvard University Press, 1987).

28. Thomas argues that "it was above all the vast expansion in the size of the known world which was causing wise men to think differently"; *Man and the Natural World* (Oxford: Oxford University Press, 1996), 167. Anita Guerrini also accepts the disappearance of animals from the urban space as an explanation for the emerging sentiments toward them, *Experimenting with Humans and Animals: From Galen to Animal Rights* (Baltimore: Johns Hopkins University Press, 2003), 75.

29. James Turner, *Reckoning with the Beast: Animals, Pain, and Humanity in the Victorian Mind* (Baltimore: Johns Hopkins University Press, 1980), 27; Guerrini, *Experimenting with Humans and Animals*, 75.

30. Rebecca J. H. Woods, *The Herds Shot Round the World: Native Breeds and the British Empire, 1800–1900* (Chapel Hill: University of North Carolina Press, 2017); Chris Otter, "Civilizing Slaughter: The Development of the British Public Abattoir, 1850–1910," in *Meat, Modernity, and the Rise of the Slaughterhouse*, ed. Paula Young Lee (Durham: University of New Hampshire Press, 2008), 89–106. Emily Pawley, "The Point of Perfection: Cattle Portraiture, Bloodlines, and the Meaning of Breeding, 1760–1860," *Journal of the Early Republic* 36, no. 1 (February 2016): 37–72. For earlier use, see Donna Landry, *Noble Brutes: How Eastern Horses Transformed English Culture* (Baltimore: Johns Hopkins University Press, 2008).

31. Abigail Woods et al., "Introduction: Centring Animals Within Medical History," in *Animals and the Shaping of Modern Medicine: One Health and Its Histories*, ed. Abigail Woods et al. (New York: Palgrave Macmillan, 2018), 5.

as the starting point for this inquiry, when the central issue of contention was no longer whether to regulate vivisection, but in what way. The leading question here would be how (rather than why) the humanitarian mindset was translated into codes of behavior in a field of science that was increasingly reliant upon animal use.[32]

This line of inquiry reflects a scholarly interest in the history of Victorian administration. Debates over the degree and nature of state intervention in nineteenth-century Britain go back to A. V. Dicey's influential 1905 essay "Law and Public Opinion in England."[33] Dicey's arguments, and in particular his claim that the mid-nineteenth century was an era of thriving individualism, were challenged by succeeding analysts.[34] One strand in the debate over the history of British governance centered on the attempt to understand the forces that initiated legislative changes. Reacting to Dicey's emphasis on theories and political ideals, Oliver MacDonagh suggested a model for government growth based on the dynamic between bureaucrats and the public.[35] Subsequent publications focused on the nuts and bolts of nineteenth-century administrative apparatuses, encompassing a wide range of fields and activities such as factories, schools, explosives production, and nature conservation.[36]

---

32. Donna Haraway grapples with the responsibilities and intimacy of animal research in *When Species Meet* (Minneapolis: University of Minnesota Press, 2007), 69–93. Ethnographic work further untangles the emotional ties between animal technicians and laboratory animals. Tora Holmberg, "Mortal Love: Care Practices in Animal Experimentation," *Feminist Theory* 12, no. 2 (August 2011): 147–63.

33. A. V. Dicey, *Lectures on the Relation between Law and Public Opinion in England* (Indianapolis: Liberty Fund Inc., 2008).

34. Henry Parris, "The Nineteenth-Century Revolution in Government: A Reappraisal Reappraised," *Historical Journal* 3, no. 1 (January 1960): 26. Peter Bartrip, who asserts that state intervention was limited almost all through the nineteenth century, agrees that in the 1890s regulation had a marked impact on public life. P. W. J. Bartrip, "State Intervention in Mid-Nineteenth Century Britain: Fact or Fiction?," *Journal of British Studies* 23, no. 1 (October 1983): 82. For the historiographical debate on state intervention in the American context, see William J. Novak, *The People's Welfare: Law and Regulation in Nineteenth-Century America* (Chapel Hill: University of North Carolina Press, 1996); Daniel Ernst, *Tocqueville's Nightmare: The Administrative State Emerges in America, 1900–1940* (Oxford: Oxford University Press, 2014).

35. Oliver MacDonagh, "The Nineteenth-Century Revolution in Government: A Reappraisal," *The Historical Journal* 1, no. 1 (January 1958): 52–67. For an analysis of MacDonagh's influence on related debates, see Valerie Cromwell, "Interpretations of Nineteenth-Century Administration: An Analysis," *Victorian Studies* 9, no. 3 (March 1966): 245–55; Catherine Mills, "The Emergence of Statutory Hygiene Precautions in the British Mining Industries, 1890–1914," *Historical Journal* 51, no. 1 (March 2008): 145–68.

36. Steven J. Novak, "Professionalism and Bureaucracy: English Doctors and the Victorian Public Health Administration," *Journal of Social History* 6, no. 4 (July 1973): 440–62; Tom Crook, "Sanitary Inspection and the Public Sphere in Late Victorian and Edwardian Britain: A Case Study in

However, the Vivisection Act was given little attention by those examining the growth of governance in nineteenth-century Britain. There are several possible reasons for this omission. In particular, the Act differed from other legislation in that it did not express a direct concern with public safety or economic interests; instead, it focused on the welfare of animals. Individual rights were not a factor in the debates about state intervention in animal experimentation, and the main objection of opponents of the legislation concerned the regulation hindering the advancement of medicine as a common good.[37] In addition, the administrative body charged with implementing the Act was markedly smaller than some of the other inspectorates; for the larger part of its first three decades, it comprised only three inspectors—one responsible for England, one for Scotland, and one for Ireland.[38] Nevertheless, the Act played a critical role in shaping British medical science and, more than other Home Office inspectorates, it provides a fruitful case study of nineteenth-century state intervention in scientific research.

Bureaucratic records not only testify to the scope of control of Victorian governance over the medical professions but also are useful for researching the history of affects and emotions. Home Office records of the Vivisection Act demonstrate what happens when care for nonhuman animals is regulated, and empathy becomes a legal demand. As the historian Janet Browne explains, since the only things available to historians are the representations of feelings, research is "turning towards the places and 'performances' sites of these representations."[39] Historians have examined Victorian experimenters' attempts at overcoming or redirecting their emotions toward the animals they vivisected.[40] In general, emotions of any sort—of human experimenters and nonhuman experimental subjects—were seen by late nineteenth- and early twentieth-century scientists as

Liberal Governance," *Social History* 32, no. 4 (November 2007): 369–93; Jill Pellew, *The Home Office, 1848–1914, from Clerks to Bureaucrats* (Rutherford, NJ: Fairleigh Dickinson University Press, 1982), 123; Roy M. MacLeod, "Government and Resource Conservation: The Salmon Acts Administration, 1860–1886," *Journal of British Studies* 7, no. 2 (May 1968): 114–50.

37. Ritvo, *Animal Estate*, 165; Richard D. French, *Antivivisection and Medical Science in Victorian Society* (Princeton, NJ: Princeton Legal Library, 1975), 103.

38. Compare, for example, the factory inspectorate, which in 1878 constituted a chief inspector, five superintendent inspectors, thirty-nine inspectors, and ten junior inspectors; Pellew, *The Home Office*, 151.

39. Janet Browne, foreword to *Medicine, Emotion and Disease 1700–1950*, ed. Fay Bound Alberti (New York: Palgrave Macmillan, 2006), ix.

40. Paul White, "Sympathy under the Knife: Experimentation and Emotion in Late Victorian Medicine," in Alberti, *Medicine, Emotion and Disease*, 100–24; Rob Boddice, "Species of Compassion: Aesthetics, Anaesthetics, and Pain in the Physiological Laboratory," *19: Interdisciplinary Studies in the Long Nineteenth Century* (June 2012): 15.

something to be curbed or locked out of the physiological laboratory.[41] Gendered constraints on the entrance of women to the laboratory were added to other markers of gentlemanly dominion, such as emotional detachment, to construct the experimental space as an emotion-free men's club.[42] However, experimenters strategically acknowledged their feeling of care toward animals when confronted with accusations of acquired emotional apathy by antivivisectionists. Tenderness toward their animal subjects was seen as acceptable and was even valued as a marker of the physiologists' refined character, but only when the emotions were moderate. At the same time as insisting that they were compassionate to animals, physiologists ridiculed their opponents' excessive sensitivity and feminine qualities.[43]

This book is about people and their efforts to comprehend the animals they used for biomedical research, and it does not attempt to avoid the anthropocentric perspective.[44] To the contrary, I delve into the anthropocentric logic at the heart of the Vivisection Act and examine how two of Western society's most celebrated institutions, law and science, coproduced knowledge about, and ethical obligations toward, nonhuman animals in late nineteenth-century Britain. Only rarely, and through the perspective of human actors, do the Vivisection Act archives reveal something about the life and suffering of nonhuman animals in British laboratories.

Feminist theorists in particular consider biology to be political discourse, the product of historical circumstances and social values, where animals are bearers of myriad cultural meanings. These scholars also carefully attend to the concrete trajectories that turned nonhuman creatures into subjects of research

---

41. Otniel E. Dror, "The Affect of Experiment: The Turn to Emotions in Anglo-American Physiology, 1900–1940," *Isis* 90, no. 2 (June 1999): 205–37; Otniel E. Dror, "Techniques of the Brain and the Paradox of Emotions, 1880–1930," *Science in Context* 14, no. 4 (December 2001): 643–60. Paul White, "Introduction," *Isis* 100, no. 4 (December 2009): 792–97.

42. Lynda I. A. Birke, *Feminism, Animals and Science: The Naming of the Shrew* (Buckingham: Open University Press, 1994).

43. A. W. H. Bates, *Vivisection, Virtue, and the Law in the Nineteenth Century* (New York: Palgrave Macmillan, 2017); Mary Ann Elston, "Women and Anti-Vivisection in Victorian England, 1870–1900," in *Vivisection in Historical Perspective*, ed. Nicolaas A. Rupke (London: Routledge, 1990); Susan E. Lederer, "Moral Sensibility and Medical Science: Gender, Animal Experimentation, and the Doctor—Patient Relationship," in *The Empathic Practitioner: Empathy, Gender, and Medicine*, ed. Ellen More (New Brunswick, NJ: Rutgers University Press, 1994).

44. Cary Wolfe, *What Is Posthumanism?* (Minneapolis: University of Minnesota Press, 2009); Anat Pick, *Creaturely Poetics: Animality and Vulnerability in Literature and Film* (New York: Columbia University Press, 2011); Irus Braverman, "Introduction: Lively Legalities," in *Animals, Biopolitics, Law: Lively Legalities*, ed. Irus Braverman (New York: Routledge, 2015), 3, 13; Kristen Stilt, "Law," in *Critical Terms for Animal Studies*, ed. Lori Gruen (Chicago: University of Chicago Press, 2018), 207.

and emblems of values.[45] The challenge of accounting for the life experiences of actual animals poses a dilemma for animal studies scholars. While some struggle to produce such accounts, other voices are skeptical as to the viability of this project, given the limitations of man-made archives.[46] The historian Ian Miller warns that giving "thoughtful agency" to animals would be "a disservice since it would invariably be an act of dominance, no matter how well intentioned."[47] Sharing this reluctance, Harriet Ritvo contemplates that "it might be more respectful to acknowledge [nonhuman animals'] inscrutability."[48] In what follows, this inscrutability of animals will be contrasted with the efforts of scientists and bureaucrats to know more about animal bodies. The administrative and scientific records will randomly provide evidence about individual creatures: a female dog who lived with her litter in a laboratory, a monkey who liked sitting in front of a fire, or an old photo of the ears of a hare. But more than anything, these records tell us something about the people who authored them.

The laboratory animal was the offspring of the conjunction of law and science. It was, in the words of the historian Paul White, "another kind of animal, whose emergence is linked to the advent of biology and physiology as professional disciplines and the rise of the laboratory as the site of knowledge production in the life sciences."[49] One of the earliest uses of the term "laboratory animal" appeared in an 1876 *Nature* report about the newly enacted British statute.[50] The term joined the existing "experimental animal," and was increasingly

---

45. Donna Haraway, *Primate Visions: Gender, Race, and Nature in the World of Modern Science* (New York: Routledge, 1990); Sarah Franklin, *Dolly Mixtures: The Remaking of Genealogy* (Durham: Duke University Press, 2007).

46. Etienne Benson, "Animal Writes: Historiography, Disciplinarity, and the Animal Trace," in *Making Animal Meaning*, ed. Linda Kalof and Georgina M. Montgomery (East Lansing: Michigan State University Press, 2011), 3–18; Erica Fudge, "What Was It Like to Be a Cow?," in *The Oxford Handbook of Animal Studies*, ed. Linda Kalof (2017), 258–78; Eben Kirksey and Stefan Helmreich, "The Emergence of Multispecies Ethnography," *Cultural Anthropology* 25, no. 4 (November 1, 2010): 545–76; Samuel Alberti, ed., *The Afterlives of Animals: A Museum Menagerie* (Charlottesville: ty of Virginia Press, 2011); Tamar Novick, "All about Stavit: A Beastly Biography," *Theory and Criticism* 51, 61–86 (in Hebrew). Haraway's *When Species Meet* is an example of a historian of science reflecting on science, technology, and concrete animals, including those used in research. Donna J. Haraway, *When Species Meet* (Minneapolis: University of Minnesota Press, 2007).

47. Ian Jared Miller, *The Nature of the Beasts: Empire and Exhibition at the Tokyo Imperial Zoo* (Berkeley: University of California Press, 2013), 14.

48. Harriet Ritvo, "Among Animals," *Environment and History* 20, no. 4 (2014): 497.

49. Paul S. White, "The Experimental Animal in Victorian Britain," in *Thinking with Animals: New Perspectives on Anthropomorphism*, ed. Lorraine Daston and Gregg Mitman (New York: Columbia University Press, 2005), 59.

50. "The Cruelty to Animals Bill," *Nature* 14 (1876): 87–88.

disseminated toward the turn of the twentieth century, only to be replaced by the contemporary "model animal" or "model organism."[51]

The Act joined physiological research in rendering animals into laboratory tools.[52] For example, nineteenth-century vivisectors chose to experiment on specific organisms over others for a variety of reasons, including availability, cost, quick payoff, the theory with which an organism was associated, and symbolic value.[53] The Act helped to shape these preferences by requiring special certificates for using dogs, cats, and some other selected species, among other requirements. In 1893, for example, the Home Office suspended the physiologist William Findlay's certificate to experiment on dogs and instructed him to experiment first on rodents.[54] The mandatory use of anesthesia in experiments, the standards of animal housing, and the reports scientists were asked to submit to the Home Office are only a few of the ways in which the law actively shaped the practices that produced the laboratory animal.

Focused on the administration of the Vivisection Act, my analysis draws extensively on the abundant Home Office memos and letters meticulously filed at the National Archives in Kew, England. I examine internal reports and memorandums composed by Home Office officials at all levels—inspectors, undersecretaries, and Home Secretaries, who were often personally involved in the Act. I make wide use of outgoing letters from the Home Office to scientists and other parties that were relevant to the implementation of the Act. The files of the Royal Commission on Vivisection from 1875 and 1906–12 are a reservoir of correspondence, pamphlets, petitions, and memos regarding the law of animal experimentation, and provide plenty of evidence on the way law and science jointly constructed the meaning of animal pain. Another leading resource for my research was the reports and memorandums written by scientific societies and leading figures in the scientific lobby, the bulk of which, in addition to a valuable

---

51. For the key characteristics of model organisms and their arguable difference from other experimental organisms, see Rachel A. Ankeny and Sabina Leonelli, "What's So Special about Model Organisms?," *Studies in History and Philosophy of Science Part A* 42, no. 2 (June 2011): 313–23.

52. In her research about the Norwegian parliament, Kristin Asdal calls to consider lawmakers in the analysis of experimental animals. Kristin Asdal, "Subjected to Parliament: The Laboratory of Experimental Medicine and the Animal Body," *Social Studies of Science* 38, no. 6 (2008): 901. The concept of rendering is borrowed from Michael E. Lynch, "Sacrifice and the Transformation of the Animal Body into a Scientific Object: Laboratory Culture and Ritual Practice in the Neurosciences," *Social Studies of Science* 18, no. 2 (1988): 265–89.

53. Barbara A. Kimmelman, "Organisms and Interests in Scientific Research: R. A. Emerson's Claims for the Unique Contributions of Agricultural Genetics," in *The Right Tools for the Job*, 200; Logan, *Before There Were Standards*, 339; Jim Endersby, *A Guinea Pig's History of Biology: The Plants and Animals Who Taught Us the Facts of Life* (New York: Arrow Books, 2008), 25.

54. Memo, May 8, 1893, HO 144/349/B14298.

file of court protocols, are held at the Wellcome Collection, London. Published primary sources include mostly parliamentary debates, parliamentary reports, and publications from both sides of the vivisection debate: scientific journals, chiefly the *British Medical Journal*, and antivivisection magazines, notably the *Zoophilist and Animal Lover*.

Chapter Outline

This book examines how British scientists and state administrators negotiated, and jointly formed, the late nineteenth-century notion of animal pain. While doing so, they recreated ideas and practices of animal experimentation, making the various understandings of pain and experiment often inextricable.

Chapter 1 depicts the medical and legal landscapes of the Act and introduces its main provisions. The chapter outlines the introduction of modern experimental physiology in Britain in the 1830s and its rise in the 1870s. From its outset, animal experimentation was confronted with public critique. The oppositional voices increased as the practice became more widespread, and were amplified by a broader suspicion toward the medical profession. The regulation of animal experimentation was also a product of developments in the legal sphere. The Vivisection Act was situated at the convergence of two trends in nineteenth-century British law—legislation on cruelty to animals and case law and the establishment of inspectorates under the Home Office as part of deeper processes of government growth and "'professionalization' of government."[55] The chapter situates the Act within the larger context of the Victorian governance and shows that the regulation of vivisection was modeled after an evolving structure that relied upon licensing and inspection. For this administrative model to work well, it was necessary for the state to ensure that the governed communities would cooperate. Therefore, the Home Office was invested in educating and disciplining researchers. The administrative structure of the Act generated intense communication between civil servants and scientists, which meant, in addition to the development of personal ties between bureaucrats and scientists, that it was the Home Office's implementation of the Act that shaped it and filled it with meaning.

Furthermore, the Act was part of a growing sphere of animal jurisprudence. The first legislation that made cruelty to animals an offense was enacted in 1822,

---

55. Roy MacLeod, introduction to *Government and Expertise: Specialists, Administrators and Professionals, 1860–1919*, ed. Roy MacLeod (Cambridge: Cambridge University Press, 2003), 6.

and it was followed by dozens of laws and amendments related to the well-being of animals. Court decisions from the nineteenth century demonstrate how animal pain became a central issue of contention in cases concerning cruelty to animals. Various experts were called upon to testify about markers of pain in different species. I trace the development of the unnecessary suffering doctrine within criminal cases of cruelty to animals, a calculation weighing degrees of pain vis-à-vis human benefits. Even though the drafters of the Act and policymakers never said so explicitly, they had adopted the utilitarian logic of this approach.

The next two chapters show that the vision of science without suffering was soon to be questioned. Brain and inoculation research, considered negligible by drafters of the Act, expanded soon after its legislation in the 1880s and 1890s. A living and active animal was essential for most experiments in these emerging sciences. The need for conscious animals challenged the widely held belief that anesthesia was the solution to all the moral difficulties with vivisection.

Chapter 2 examines debates about whether the observation of animals that had been mutilated by chirurgical means should be considered a type of experiment to which the Act did not apply. The chapter opens with a description of the state of knowledge about pain reduction in the nineteenth century, together with the social and cultural meanings of pain in Britain. As a result of Victorians' growing aversion to physical pain and the production of sedatives, anesthesia became a dominant technology that impacted animal experimentation, the perception of the morality of vivisection, and its regulation. Shortly after the Act came into force, professor of physiology at King's College Gerald Yeo planned to examine the impact of a certain injury on dogs. Destroying the dogs while sedated, as the Act instructed, would have nullified Yeo's aims of observing the impact of the injury on the dogs' faculties. Yeo, therefore, applied to the Home Office for a certificate that was needed in cases when an animal was allowed to recover from anesthesia. However, the Home Office asked him to supplement his application with another certificate, mandatory for experimenting on dogs without anesthesia. A disagreement over the classification of Yeo's planned observation then followed. Yeo refused to ask for a special certificate to experiment without anesthesia on dogs, probably wishing to avoid being labeled as one who conducted cruel experiments on this popular species. His argument was the following: he agreed that the first part of his plan—surgery—came under the Act and was subject to its requirements. However, he challenged the Home Office over the question of whether the following step, that is, observation of the dogs, should be considered an experiment. Yeo contended that his planned observation of the dogs was not an experiment, and therefore no additional certification was needed.

Gerald Yeo and the Home Office were engaged in an escalating correspondence when, in November 1881, the Victoria Street Society for the Protection of Animals from Vivisection initiated a prosecution against neurologist David Ferrier, raising very similar issues. The event that brought about the prosecution had taken place at the International Medical Congress in London. During the meeting, Ferrier had confronted German physiologist Friedrich Goltz on the latter's dismissal of brain localization theory. An informal gathering in the physiological laboratory of King's College followed the rival neurologists' speeches. Ferrier then presented the audience with two monkeys, one partially paralyzed and the other deaf, as living proofs of the ability to control specific bodily functions through particular areas in the brain. In inner scientific circles, the presentation was a great success, but for the Victoria Street Society, it was bluntly illegal.

The leading antivivisection organization argued in court that the paralyzed monkey and the deaf monkey were suffering in an ongoing experiment. The society claimed that Ferrier had conducted experiments on monkeys without a license and thus violated the Act. The strategy of the defense was to separate the surgical event that damaged the monkeys' brains from their observation thereafter. Ferrier's lawyers thus argued that the surgical event was past the Act's limitation period at the time the prosecution was filed and that the latter event did not constitute an experiment under the Act. Ferrier, according to this reasoning, did not have to hold a license, or to follow any of the Act's provisions for that matter. The defense's framing of Ferrier's presentation of the mutilated monkeys convinced the court. The judge dismissed the summons against him, not without lamenting the absence of a clear legal definition for 'experiment.'

The widely reported trial affected all sides: the hopes of the Victoria Street Society regarding the Act, however minimal to begin with, were dashed, and although the trial ended with a victory for the scientific lobby, the fear of prosecution resonated within the community. The Home Office, which was not party to the legal dispute, reconsidered its policies regarding brain experiments. From then on, civil servants were alarmed when handed applications to allow animals to recover from anesthesia, especially when animals that the public favored, such as dogs or monkeys, were involved.

What is the definition of an experiment? When does it start, and when does it end? During the debate between Yeo and the Home Office, questions that touched upon the foundations of scientific ethos reached the bureaucratic domain. The same questions were rearticulated by litigators and set before a judge in Ferrier's trial. State administrators and lawyers were to provide their interpretation of the term 'experiment,' which consequently obliged physiologists to follow the Act's requirements or exempted them from it. At stake, therefore, were the routines of experimental practice as well as the boundaries of the law—and the two were mutually dependent. The disputes surrounding the administrative classification of brain research were chiefly about specifying the reach of the law, but they were also about expertise and access to knowledge. A book published incognito by Yeo in 1883 disclosed this territorial as well as epistemic conflict. In this provivisection essay, Yeo set out arguments against a lay interpretation of animal behavior, such as confusing reflex movements and manifestations of suffering, stressing the privileged access of physiologists to knowledge about animal pain instead. The competing claims for the authority to interpret animal pain would continually inform the communication between the Home Office and scientists regarding the Act.

Chapter 3 examines the challenges posed by the bureaucratic classification of inoculation experiments and serum production, where the emphasis shifted from the question of what constituted an experiment to what constituted pain. The word vivisection is drawn from Latin, and it means cutting alive (*vīvus* living + *sectio* cutting). The term was in popular use in nineteenth-century Britain, and although some argued that it failed to reflect the employment of anesthesia, what it denotes correlated with the mainstream of physiological practice up to the 1870s. British legislators had in mind mainly surgical operations on living organisms when they drafted the Act; the mandatory use of anesthesia, for example, and the requirement to destroy animals before their recovery, were tailored for these kinds of experimentation.

In the 1880s inoculation took over the British medical research scene. This challenged scientists and civil servants alike with the question of whether inoculation studies and serum production constituted an experiment under the Act. Even more puzzling was the question of whether these studies were calculated to give pain to living animals. The phenomenology of inoculation research was fundamentally different from the common image of physiology: the symptoms following inoculation appeared only gradually and with different intensity. Sometimes no symptoms developed at all. Additionally, for many Victorians, disease as a natural occurrence was considered an inconvenience at most. Hence, even when the symptoms developed into a full-blown disease, some contended that the kind of pain involved was not what the law intended to alleviate.

The civil servants at the Home Office were therefore perplexed when, in 1879, William Smith Greenfield, the director of the Brown Animal Sanatory Institution, asked if he needed a certificate exempting him from the use of anesthesia to experiment with inoculation. The first to process the inquiry, inspector George Busk, thought that inoculation research was not an experiment, and hence it did not require any certification. Assistant Undersecretary of State

Godfrey Lushington, however, was not sure. Following his recommendation, the Home Office forwarded the inquiry to the Queen's Law Officers. The Solicitor General and Attorney General adopted a broad interpretation of the Act and opined that inoculation research was indeed an experiment. The chapter analyzes five more cases in which the Home Office outsourced decisions regarding the legal status of inoculation research and serum production to the Law Officers. Time after time, the Law Officers on duty had to tackle the question of whether inoculation experiments came under the Act: Were the experiments calculated to give pain? First, they had to decide whether the operation under consideration was an experiment at all and, second, whether it was painful. An affirmative reply to both questions would require the experimenters to hold a license and possibly a special certificate, and that the places of experimentation be registered in the Home Office records.

While making sense of the terms pain and experiment, civil servants and physiologists negotiated their field of expertise, clashed over authority, and translated epistemic questions about suffering into bureaucratic formulas. However, the map of rivalries over the inclusion of inoculation operations in the Act did not square with the 'state versus scientists' narrative that characterized the brain research debates. Unsurprisingly, pathologists preferred that the law should not restrict inoculation research. Neither was the administration at the Home Office anxious to expand its areas of responsibility to cover new kinds of experiments. In addition to the challenge of identifying and quantifying the pain of disease, the Home Office was troubled by the impact the decision might have on its workload. Furthermore, such expansion of its authority was likely to lead to a public relations crisis; including inoculations and serum production in the official records would have resulted in an increase in reported experiments, which might draw undesired public attention. Internal memos also express a concern within the administration regarding its own role in fostering the progress of science.

The cases of inoculation and antitoxin production illuminate the volatility of the concept of pain and how the Act's bureaucracy tied it into the definition of an experiment. In chapter 4 our focus moves from governmental offices to physiology laboratories, and from abstract ideas about pain to hands-on examination of animals by vivisection inspectors. The chapter follows the day-to-day activities involved in the implementation of the Act. Sent to examine registered laboratories, inspectors had to record their impressions of the animals they examined and to recognize—or dismiss—symptoms of pain. The chapter explores the relations between law and science, regulation, and empathy, through the visitation notes taken by three principal inspectors employed between 1876 and 1912:

George Vivian Poore, James Alexander Russell, and George D. Thane. Their notes from 1894, 1895, 1897, and 1899 provide a rare glimpse into the enforcement of the Vivisection Act. These notebooks bring a fresh perspective to late nineteenth-century bureaucracy as they provide an informal record of the inspectors' activities, and, together with their memos and correspondence, afford a sense of the everyday practice of the law.

Qualified witnessing has been recognized by science studies scholars as fundamental to modern science. I borrow this insight about witnessing being constitutive of empirical science to explain that inspectors' role in the laboratories was greater than being bureaucratic watchdogs. Their work, and particularly their laboratory inspection, were significant in incorporating the ethical norm set by the Act into the research carried out in laboratories. They educated experimenters about the Act's requirements, and formally and informally advised physiologists how to coordinate their practice with the law.

Trained in the medical professions, the inspectors were at home in both the scientific and the bureaucratic worlds, and became agents of coproduction. For the inspectors, who were constantly asked to respond to questions concerning animals' pain, implementing the law was inherently intertwined with an experience of empathy. While examples of their role in shaping the interpretation of pain through consultation with Home Office officials are scattered throughout this book, chapter 4 focuses specifically on laboratory visits. It examines inspectors' testimonies about the animals they saw and held during their visits to registered places, analyzes their depictions of the animals, and asks what indicators of pain they were looking for. The inspectors were reluctant to report animal suffering in laboratories even though they closely observed numerous diseased and injured animals. I suggest that their alleged blindness to animal suffering was due, among other things, to the difficulty of defining pain. They focused instead on procedural breaches of the Act and were proactive in educating the experimenters about the legal requirements.

While the inspectors were hesitant to report to their supervisors at the Home Office about suffering animals in the laboratories they visited, they were nonetheless eager to educate the physiologists and supervise their compliance with the Act. Moving between the legal and scientific spheres, they played a central role in shaping administrative policies toward animal experimentation and had a significant influence on the work of physiologists, exemplified among others in the design of the experimental space. When inspectors guided experimenters to modify their practices or redesign animal shelters in accordance with the regulations for the prevention of cruelty to animals, they intertwined practical and normative aspects of animal research.

The role of law in the coproduction of normative order and physiological knowledge was epitomized by the inspectors acting as witnesses to live experiments. Calls to include them as supervisors in all experiments were dismissed on practical grounds, but there are records of inspectors attending experiments by order of the Home Office, or coincidentally, while inspecting registered places. Starting with the Act and under the attentive gaze of the inspectors, experimental laboratories generated not only facts but also an ethically acceptable science.

The antivivisection movement, which played a pivotal role in the history of animal welfare in Britain in general and in the Victorian period in particular is admittedly marginal in this book's narrative. Antivivisectionists are mentioned only to the extent that they were directly involved in issues related to the administration of the Act and the question of animal pain. This is because there exists splendid historical research on the antivivisection movement, exploring the social and cultural contexts of its evolvement and explaining its appeal to specific sections of the British public.[56] Notwithstanding the decision to focus on bureaucrats and scientists, the next chapter is built around one such antivivisection activist.[57]

Chapter 5 opens with the events leading physiologist William Bayliss to file, in 1903, for libel action against Stephen Coleridge, honorary secretary of the National Anti-Vivisection Society, for reciting an account written by Swedish antivivisection activist Louise Lind-af-Hageby about Bayliss's demonstration on a dog. Historians have already examined the social circumstances that surrounded what became known as the brown dog affair, but this chapter zooms in on the legal proceedings. I use hitherto unexamined material to reveal the role of the Home Office in the case and examine the informal investigation that they carried out in parallel to the public hearings. Bayliss won the case, but in a way, Lind-af-Hageby's cause was vindicated. In light of the new public interest in the operations of laboratories, a second Royal Commission reexamined the law in the years 1906–12. Unlike the first Royal Commission on Vivisection of 1875, which led to the legislation of the Act, the second Royal Commission has received little, if any, scholarly attention. The Home Office was a central source

56. E.g., Bates, *Vivisection, Virtue, and the Law*; French, *Antivivisection and Medical Science*; Susan Hamilton, *Animal Welfare & Anti-Vivisection 1870–1910: Frances Power Cobbe* (London: Taylor & Francis, 2004); Hilda Kean, *Animal Rights: Political and Social Change in Britain since 1800* (London: Reaktion Books, 1998); Coral Lansbury, *The Old Brown Dog: Women, Workers, and Vivisection in Edwardian England* (Madison: University of Wisconsin Press, 1985); Nicolaas Rupke, *Vivisection in Historical Perspective* (New York: Routledge, 1990).

57. A version of this chapter was published in Shira Shmuely, "Law and the Laboratory: The British Vivisection Inspectorate in the 1890s," *Law & Social Inquiry* 46, no. 4 (November 2021): 933–63.

of information for the commissioners not only about the law but also about the state of physiological practice. At the Commission, the inspectors served as experts in anything from the number of experimenters active in Britain, through changing research trends, to providing knowledge about animal sentience. The Commission laid special emphasis on pain, and, while attempting to classify various pain-inducing operations, it often used human–nonhuman analogies.

The chapter ends with a discussion of a 1912 libel lawsuit filed by Lind-af-Hageby against a few people involved in a *Pall Mall Gazette* publication who scorned an antivivisection campaign she had sponsored. Lind-af-Hageby argued before the King's Bench Division of the High Court of Justice that the defenders had damaged her reputation when claiming, in the *Pall Mall Gazette*, that she had distributed misleading material about vivisection. The defenders, however, contended that their portrayal of Lind-af-Hageby's campaign was accurate and that she was misrepresenting animal experimentation in her pamphlets and posters. Here the Act appeared as a defense claim: the defenders reclaimed the law to argue that, since British medical research was performed in accordance with the requirements of the Vivisection Act, the pain inflicted was minimal. The defense further claimed that because the Act was in force, an attack on vivisection was an attack on English law. Despite remarks to the contrary, both parties used the case to make grander claims about the legitimacy of vivisection.

The outbreak of the First World War reshuffled British politics and marginalized the antivivisection campaign. The Vivisection Act remained substantially unaltered for more than a century, while the number of experimental animals in Britain surged to millions. However, the Royal Society for the Prevention of Cruelty to Animals revived public interest in the Act with its critique of the Act's administration in the early 1960s, a move that instigated the repeal and replacement of the Act in 1986. Chapter 6 provides an overview of the main developments in animal experimentation law in twentieth-century Britain. It shows how the efforts to identify and categorize animal pain, which had originated in the apparatuses of implementing the Act, became a central medico-legal task. New scientific knowledge about animal sentient capacities led to a reconsideration of the key element in the Act, the definition of pain, and particularly that of the species that were to be protected in light of their sentient capacities. At the same time, the state bureaucracy introduced increasingly sophisticated schemes to quantify and classify animal pain in laboratories. The book concludes with contemporary debates about animal ethics that both replicate late nineteenth-century concerns and present a more nuanced picture of the role of pain in accounting for nonhuman lives.

1

THE LEGAL AND SCIENTIFIC LANDSCAPES OF THE ACT

The Vivisection Act was an important milestone in the histories of physiology and of animal law. Decades-long processes in these distinct domains coalesced in the administrative system that governed "any experiment calculated to give pain." Transformations in the scientific understanding of pain and the introduction of new methods of desensitization overlapped with the demise in the theological significance of suffering. At the same time, new anticruelty laws expressed emerging care for animal welfare, a sentiment that would become a matter of national pride. This chapter first sets the legal and medical scene into which the Act was introduced. The following sections tell the Act's legislative history and lay out its main provisions, before delving, in the rest of this book, into the details of the law's implementation, its evolving routines, and the disagreements and settlements regarding animal pain that gave rise to the bureaucracy of empathy.

A Benthamite Blueprint

The Vivisection Act's governing style, legal logic, and ethical commitments reflected nineteenth-century reformist ideals as devised by the philosopher Jeremy Bentham (1748–1832).[1] The primary reason was that it embodied ideas of minimizing suffering and calculating the necessity of pain infliction in relation to its potential benefits. The secondary reason was that the Act was designed

---

1. Pellew, *The Home Office*, 4. About Bentham's legacy in nineteenth-century medical reform, see Adrian J. Desmond, *The Politics of Evolution: Morphology, Medicine, and Reform in Radical London* (Chicago: University of Chicago Press, 1989), 26–27; Michael Brown, "Medicine, Reform and the 'End' of Charity in Early Nineteenth-Century England," *English Historical Review* 124, no. 511 (December 2009): 1370. Nevertheless, the influence of Bentham's political philosophy on nineteenth-century administrative reform is debatable, see an overview in L. J. Hume, "Jeremy Bentham and the Nineteenth-Century Revolution in Government," *Historical Journal* 10, no. 3 (1967): 361–62. For an example of challenging the influence of Bentham, see David Roberts, "Jeremy Bentham and the Victorian Administrative State," *Victorian Studies* 2, no. 3 (1959): 206.

as an administrative system and did not follow the prohibitive model of cruelty to animals laws. The regulation of vivisection was a Benthamite project in form and content. Bentham developed his theory of law as a critique of William Blackstone's *Commentaries on the Laws of England* (1765–69). In contrast to Blackstone's collection of customs and precedents that derive their legitimacy from natural law, Bentham aspired to reinvent and codify English law according to reason and a clearly stated principle: to produce the greatest happiness to the highest number. Known for his fondness for neologisms, Bentham termed his theory "utilitarianism."

In *Introduction to the Principles of Morals and Legislation* (1789), Bentham defined morality in terms of pain and pleasure, articulating, "Nature has placed mankind under the governance of two sovereign masters, *pain* and *pleasure*. It is for them alone to point out what we ought to do, as well as to determine what we shall do."[2] In a note in the essay, Bentham criticized English law's failure to protect animals, with a statement that would become an axiom for twentieth-century animal welfare and rights advocates: "The question is not, Can they *reason*? Nor, can they *talk*? But, can they *suffer*?"[3] Bentham equated humans' and nonhuman animals' vulnerability to pain, arguing that since minimizing pain and maximizing happiness were the principal considerations in utilitarianism, animals' interest in avoiding pain should be considered by lawmakers. In the same note, Bentham justified the slaughtering of animals for food for its contribution to human happiness but did not discuss animal experimentation.

Politics and medicine were interrelated in Bentham's thought. Laws were justified as long as they contributed to the welfare of the people, and good health was immanent to this objective. This philosophy gave rise to what the historian Cathy Gere identifies as an era of medical utilitarianism, in which the task of balancing pleasure and pain was intertwined with politics and medicine. As Gere explains, "the measure of a good or bad law was how much pain or pleasure it caused, and these effects were mediated by medical facts about physical vigor."[4]

Bentham did not live long enough to tackle the challenges that vivisection posed to utilitarian calculus, chiefly the hurdles in the making of pain into comparable medical facts. However, Bentham was personally involved with the

---

2. Jeremy Bentham, *An Introduction to the Principles of Morals and Legislation* (Mineola, NY: Dover Publications, [1789] 2007), 1.

3. According to Adrian Desmond, vivisection "upset many medical Benthamites and Dissenters, for whom care was a sacred duty and wasteful sacrifice a sin against Creation." Desmond, *The Politics of Evolution*, 189.

4. Cathy Gere, *Pain, Pleasure and the Greater Good: From Panopticon to the Skinner Box and Beyond* (Chicago: University of Chicago Press, 2017), 93.

Anatomy Act (1832) legislation, which regulated the use of human cadavers in medical research.[5] The Act aimed to guarantee the supply of corpses for British anatomists, while also annihilating the shady trade in corpses for pathological anatomy, a practice that stirred fear and unrest among its ultimate victims, poor and impoverished members of society. The Anatomy Act mandated license holding and record-keeping and established the first centrally financed and administered national inspectorate, followed a year later by the factory inspectorate.[6] It was then that the seeds of the Vivisection Act were planted in British law books.

Vivisection Professionalization

Experimental physiology had developed and gradually solidified as a distinct scientific discipline in Britain during the nineteenth century. Animal experimentation was not foreign to English science; in the mid-seventeenth century, the physician William Harvey (1578–1657) examined blood circulation in living animals. Harvey's work, which set the foundations for medical science, was so influential that his findings continued to provide central evidence for supporters of vivisection long into the nineteenth century.[7] However, it took two more centuries before figures such as Charles Bell (1774–1842), Marshall Hall (1790–1857), and a few other physiologists working independently, turned experimental physiology into a systematic, inductive pursuit in Britain. Through the destruction of different tissues of living animals, nineteenth-century physicians mapped nerves, blood vessels, and muscles. They examined the effects of substances, such as various poisons, on body functioning, and produced diseases to explore potential remedies. Submitting animals to extreme conditions such as hunger to study metabolism, and cold to explore hypothermia and temperature

---

5. French, *Anti Vivisection & Medical Science*, 145. For the legislative history of the Anatomy Act, see Richardson, *Death, Dissection, and the Destitute*, 109–215. For a re-reading of the Anatomy Act following Richardson, see Elizabeth T. Hurren, *Dying for Victorian Medicine: English Anatomy and Its Trade in the Dead Poor, c.1834–1929* (Basingstoke: Palgrave Macmillan, 2014), 20–37.

6. Ruth Richardson, *Death, Dissection, and the Destitute* (London: Routledge & Kegan Paul, 1987), 108.

7. Harvey's significance to pro-vivisection arguments is also a sign of the shortage in demonstrated achievements of the practice until the late nineteenth century. For an authoritative scholarship on the history of vivisection, see Anita Guerrini, *Experimenting with Humans and Animals: From Galen to Animal Rights* (Baltimore: Johns Hopkins University Press, 2003); Anita Guerrini, "The Ethics of Animal Experimentation in Seventeenth-Century England," *Journal of the History of Ideas* 50, no. 3 (July 1989): 391–407.

regulation, was also a common experimental and experiential practice in the century.

A radical reform movement moved through the British medical world in those years, coming to London from Paris through Edinburgh. Medical students and physicians who, like Hall, had not belonged to traditional elite institutions targeted monopolistic corporations such as London's Royal College of Physicians and the Royal College of Surgeons in their Bentham-inspired call for a revision of medical studies and practice in Britain. Their efforts contributed to the legislation of the Medical Act (1858), which established the Medical Council and, to a certain extent, democratized medical studies. The historian Adrian Desmond shows how the politics of the radical reform movement in the 1830s and 1840s were reflected in medical research substantially. Radical anatomists favored materialistic and naturalistic "philosophical anatomy," and embraced Lamarckian ideas, which explained the power of change as residing within organisms rather than directed by superior forces.

By contrast, Oxbridge and London's establishment medicine (surgeons, anatomists, and collectors) advocated for "functional anatomy," which incorporated divine design in a theory of fixity of species, thereby justifying the existing social order.[8] In this scheme, vivisection was generally associated with the modern, advanced practice of philosophical anatomy. Dissection, a practice preferred by the traditional school of functional anatomy, was deemed by reformers a failed and outdated mode of studying the ever-changing, revolutionary, natural world. Nevertheless, some medical reform groups opposed animal experimentation as part of their vision of proper medical care.[9]

The historians William Coleman and Frederic L. Holmes maintain that nineteenth-century physiology did not differ from its predecessors in modes of reasoning and observation of functions of the body. Rather, physiology changed in its intense and continuous character, which formed new institutions and collective investigative enterprises first in continental Europe and later in Britain. Chairs in physiology were established in London in the mid-1830s, a decade after Parisian institutions had established similar positions. When University College

---

8. This is a broad sketch of the turmoil in British life sciences in the nineteenth century. For a thorough analysis, see Desmond, *The Politics of Evolution*; Adrian Desmond, "Redefining the X Axis: 'Professionals,' 'Amateurs' and the Making of Mid-Victorian Biology: A Progress Report." *Journal of the History of Biology* 34, no. 1 (April 2001): 3–50; William Coleman and Frederic L. Holmes, *The Investigative Enterprise: Experimental Physiology in Nineteenth-Century Medicine* (Berkeley: University of California Press, 1988).

9. Desmond, *The Politics of Evolution,* 189.

finally decided to found a chair for physiology in 1836, it requited the anatomist and physiologist William Sharpey (1802–80) for the post.

Sharpey's student, Edward Sharpey-Schafer (1850–1935), who had undertaken to write the history of his profession of physiology, explained that "it was not possible at the time to obtain a trained physiologist to occupy it; there were none in the kingdom."[10] Even with the incorporation of physiology chairs into medical schools, the field did not take off. Historians portray 1840–70 as a period of stagnation of English physiology, which was thriving on the Continent.[11] However, in the 1870s experimental physiology had strengthened its hold on British science, partly because of changes in medical training fostered by the General Medical Council.

Unwasted Pain: Marshall Hall's and the British Medical Association's Ethical Codes

The physician and neurophysiologist Marshall Hall introduced an early attempt to adjust vivisection to a moral cost-benefit calculus through ethical guidelines. Born to a radical Methodist family in Basford, England, Hall earned his MD degree at Edinburgh University in 1812. After two years of residency in the Edinburgh Royal Infirmary, Hall traveled on the Continent for a few years, where he visited several medical schools. Back in London, Hall established a successful practice while devoting his research time to vivisection, which was spreading in continental science.[12]

As was typical in the decades preceding the Vivisection Act, Hall operated his research laboratory at home. In the comforts of his residence, he experimented with dogs, mice, hedgehogs, bats, birds, fish, turtles, toads, frogs, a donkey, and even an alligator. Charlotte, Hall's wife, remembered how "the animals required for all these investigations formed a little menagerie in a room of our house devoted to that special purpose."[13] An unexpected movement of a salamander's

---

10. E. Sharpey-Schaper, "Observations on the History of Physiology in Great Britain," *British Medical Journal* 2, no. 3747 (October 1932): 781–83.

11. Gerald L. Geison, *Michael Foster and the Cambridge School of Physiology: The Scientific Enterprise in Late Victorian Society* (Princeton, NJ: Princeton University Press, 1978), 14; Stella V. F. Butler, "Centers and Peripheries: The Development of British Physiology, 1870–1914." *Journal of the History of Biology* 21, no. 3 (October 1988): 473–500.

12. Diana E. Manuel, "Marshall Hall (1790–1857): Vivisection and the Development of Experimental Physiology," in Nicolaas Rupke, *Vivisection in Historical Perspective* (New York: Routledge, 1990) 78–104, 79; Turner, *Reckoning with the Beast*, 88.

13. Turner, *Reckoning with the Beast*, 82.

detached tail during an experiment shifted Hall's early fascination with the blood to the nervous system and specifically the reflexes.[14] Physiological questions inherent to this new field of research—animal volition, sensation, and consciousness—were echoing the moral concerns raised by vivisection's critics. Antivivisectionists, still a marginal voice, targeted Hall in the medical press.[15]

The emergence of new theories about the nervous system gave rise to the previously unimagined concept that sensation could be detached from other life faculties. Eighteenth-century understanding that all parts of the body were alive with sensation and motion was replaced with the insight that sensation and movement could be distinguished, and that feelings and emotions were linked to the neuromuscular system and specific sense organs.[16] The French physiologist François Magendie (1783–1855) and Charles Bell identified, independently, the "Bell-Magendie law," according to which the dorsal (posterior) roots of the spinal cords were primarily associated with sensibility and the ventral (anterior) roots with mobility. In this new understanding of bodily mechanisms, explains the historian Anita Guerrini, "the organism as a whole could not feel; only the nervous system could."[17]

Early nineteenth-century physiological and anatomical knowledge about the nervous system thus paved the way for the concept of an "unfeeling, yet living, body."[18] The conceptualization of pain as a set of physiological processes facilitated experimenters' ambition, including Hall's, to create what the historian Otniel Dror calls a "pain-less animal."[19] As Dror points out, the development of decerebration techniques was to provide a solution to the problem of animal

---

14. Manuel, Diana E. "Marshall Hall, F.R.S. (1790–1857): A Conspectus of His Life and Work," *Notes and Records of the Royal Society of London* 35, no. 2 (December 1980): 135–66, 145.

15. A. W. H. Bates, *Vivisection, Virtue, and the Law in the Nineteenth Century* (London: Palgrave Macmillan, 2017), https://link.springer.com/chapter/10.1057/978-1-137-55697-4_2.

16. Michael Gross, "The Lessened Locus of Feelings: A Transformation in French Physiology in the Early Nineteenth Century," *Journal of the History of Biology* 12, no. 2 (1979): 233.

17. Guerrini, *Experimenting with Humans and Animals*, 74. Examining developments in French physiology, Michael Gross shows the replacement of the eighteenth-century theory concept "that all parts of the body are alive with sensation and motion," with the theory that "feelings were felt not peripherally but in the brain only, borne to it by fibers of the nervous system," in Gross, "The Lessened Locus of Feelings," 233. Compare this to the abstraction of life with the solidifying of the discipline of biology early in 1800 and the rise of Foucauldian biopolitics, in Helmreich, *Alien Ocean*, 6. For animal pain and comparative psychology, see Liz Gray, "Body, Mind and Madness: Pain in Animals in Nineteenth-Century Comparative Psychology," in *Pain and Emotion in Modern History*, ed. Rob Boddice (London: Palgrave Macmillan UK, 2014), 148–63.

18. Stephanie J. Snow, *Operations without Pain: The Practice and Science of Anaesthesia in Victorian Britain* (Basingstoke: Palgrave Macmillan, 2005), 9.

19. Otniel E. Dror, "Techniques of the Brain and the Paradox of Emotions, 1880–1930," *Science in Context* 14, no. 4 (December 2001): 651.

pain in the laboratory; this would create the laboratory as a pain-free space, and presumably suppress critique by antivivisectionists. This is part of what motivated Hall, who, while researching reflex physiology in the 1830s, developed techniques of decerebration to manipulate sensation capacities and examine the body motions of various animals.[20]

In an 1831 essay on circulation, Hall introduced an ethical code for experimenters. Hall opened the volume lamenting that "unhappily to the physiologists, the subjects of the principal department of this science, that of animal physiology, are sentient beings; and every experiment, every new or unusual situation of such a being, is necessarily attended by pain or suffering of a bodily or mental kind."[21] Since scientific investigations were "exposed to peculiar difficulties," it needed to be "regulated by peculiar laws."[22] Hall's seven "principles" for physiological research wove together the treatment of animals, the relationship between investigators, and scientific accuracy: experiments should be done only upon necessity, and observation was preferable over experimentation on living animals; each experiment should have a distinct and definite object; experimenters should avoid unnecessary repetition; experimenters should inflict the least possible pain; experiments should be properly observed and witnessed; facts generated by experimenters should be announced to the public in a simple way; and quoting the opinions of other authors should be in their own words, to avoid the misinterpretation of others' ideas.

The pain principle included a recommendation to substitute dead for living animals when applicable. Such a switch had "the great advantage of at once avoiding the infliction of pain, and its effects in complicating the results of the experiment."[23] Likewise, the subject of the experiment should be chosen from "the lowest order of animals appropriate to our purpose, as the least sentient." Specifically, Hall recommended using "batrachian reptiles," which could be "easily deprived not only of sensation, but of motion," by the decerebration technique

20. About Marshall Hall's work in the context of nineteenth-century reflex research, see Edwin Clarke and L. S. Jacyna, "*Nineteenth-Century Origins of Neuroscientific Concepts* (Berkeley: University of California Press, 1987), 116–22.

21. Marshall Hall, *A Critical and Experimental Essay on the Circulation of the Blood* (London: R. B. Seeley and W. Burnside, 1831), B.

22. Hall, *A Critical and Experimental Essay*, 1. For experimentalism and antivivisection criticism by medical Benthamites and Dissenters in Marshall Hall's time, see Desmond, *The Politics of Evolution*, 188–92.

23. In addition to the guidelines, Hall suggested the establishment of a society for physiological research. The society's members, he offered, will pre-examine experiments and supervise them. Such a society was established forty-five years later. Hall, *A Critical and Experimental Essay*, 9.

he had developed.[24] From Hall's perspective, animals' well-being and valuable physiological research were compatible objectives. A motionless animal—dead or decapitated—was less prone to "complicate" the procedure and mess up the results.

In a paper read before the Royal Society in 1833, Hall argued, "sensation can act, in inducing muscular motion, only through the medium of volition." If volition—"the will, and not the power, to move"—is destroyed through decapitation, "the agency of sensation [is] excluded." In these circumstances, "the influence of external impressions, which might be supposed to induce pain, must have been exerted upon some property of the nervous system different from sensibility."[25] In other words, an ostensibly painful stimulus to detached limbs would generate a motion that could be erroneously understood to be a manifestation of sensation. This theory paved the way for later arguments privileging knowledgeable physiologists over lay observers in telling true pain from mere reflexive movements. Hall's investigations of neurophysiological mechanisms of sensation and movement provided him with the aspiration (and, in his view, also the capacity) to control animal pain by surgical means. Techniques of decerebration not only helped Hall develop his theory about reflexes but also provided him with a useful device to disarm vivisection critique. Hall's principles were an early exemplification of medical utilitarianism in vivisection. Variations of the necessity and the pain principles will reappear in the Vivisection Act, and importantly, the incorporation of medical technologies to render animals insentient into the normative code, thus intertwining scientific practice and moral conduct through the regulation of animal experimentation.

Public critique on vivisection had increased together with the growth of physiology, echoed by debates among medical men regarding the morality of their deeds. Four decades after Hall, the British Association for the Advancement of Science (BAAS) (established in 1831) introduced another scheme for an ethical code. In a meeting in Liverpool in September 1870, the General Committee of the BAAS instructed the Committee of Section D (Biology) to "draw up a statement of their views upon Physiological Experiments" and to consider "from time to time whether any steps can be taken by them, or by the Association, which will tend to reduce to its minimum, the suffering entailed by legitimate physiological

---

24. Hall, *A Critical and Experimental Essay*, xx. Hall's principles should also be read against the backdrop of the 1830s medical radicalism and calls for medical reform. Desmond, *The Politics of Evolution*.

25. Marshall Hall, "On the Reflex Function of the Medulla Oblongata and Medulla Spinalis," *Philosophical Transactions of the Royal Society of London* 123 (January 1833): 635–65, 642.

inquiries; or any which will have the effect of employing the influence of this Association in the discouragement of experiments which are not clearly legitimate on live animals."[26]

At the next meeting in August 1871, the committee presented the Report of the Committee Appointed to Consider the Subject of Physiological Experiments, signed by M. A. Lawson, G. M. Humphry, John H. Balfour, Arthur Gamgee, William Flower, John Burdon Sanderson, George Rollestone:

I. No experiment which can be performed under the influence of an anaesthetic ought to be done without it.

II. No painful experiment is justifiable for the mere purpose of illuminating a law or a fact already demonstrated; in other words, experimentation without the employment of anaesthesia is not a fitting exhibition for teaching purposes.

III. Whenever, for the investigation of a new truth, it is necessary to make a painful experiment, every effort should be made to ensure success, *in order that the suffering inflicted may not be wasted*. For this reason, no painful experiment ought to be performed by an unskilled person with insufficient instruments and assistance, or in places not suitable to the purpose, that is to say, anywhere except in physiological and pathological laboratories, under proper regulations.

IV. In the scientific preparation for veterinary practice, operations ought not to be performed upon living animals for the mere purpose of obtaining greater operative dexterity.[27]

The BAAS's attempt at self-regulation might have been a strategic move to reduce public critique and to prevent state intervention in experimental practices. The report was circulated among scientists, but it is unclear whether the

---

26. The British Association for the Advancement of Science, *Report of the Fortieth Meeting of the British Association for the Advancement of Science* (London: Taylor and Francis, 1871), lxii, https://www.biodiversitylibrary.org/item/93095#page/8/mode/1up. Jim Endersby explains that the specialist sections of the BAAS embodied the hierarchy of disciplines in its time of constitution. Section D (Biology) came after the Mathematical and Physical Sciences, Chemistry and Mineralogy, and Geology and Geography. It was followed by Medicine, Statistics, and Mechanical Sciences. Endersby, "Classifying Sciences," 63.

27. British Association for the Advancement of Science, *Report of the British Association for the Advancement of Science* (London: Taylor and Francis, 1872), 144, https://www.biodiversitylibrary.org/ia/reportofbritisha72brit/#page/263/mode/1up; emphasis added.

guidelines had any impact on practicing physiologists.[28] There was no mention of the report's recommendations in the first English handbook for the physiological laboratory published by John Burdon-Sanderson (1828–1905) and others in 1873.[29] The BAAS's guidelines testify that the regulation of vivisection in 1876 was not the state's unilateral imposition of medical utilitarianism; the idea of subordinating the infliction of animal pain to a cost-benefit calculus was also on the minds of individuals such as Marshall Hall and, later, organizations such as the BAAS.

Both Hall's principles and the association's guidelines tied the professional qualifications of experimenters together with the moral aim of reducing the suffering of animals in research. Both also suggested medical technologies to resolve the problem of pain in laboratories: while Hall was a firm believer in decapitation, the association's guidelines placed anesthetics as the ultimate solution to animal experimentation's moral question. The BAAS's guidelines displayed a crystalized utilitarian logic when stated, in the third recommendation of the report, that the suffering inflicted by experiments "may not be wasted." Meanwhile, the idea of a surplus pain—"wasted pain" in BAAS terms—was also evolving in the legal realm as the doctrine of "unnecessary suffering." The scientific and legal conceptualizations of surplus pain merge in the Vivisection Act.

Nineteenth-Century Animal Jurisprudence

The Vivisection Act emerged at the height of evolving animal jurisprudence that concerned, in addition to cruelty, also conservation, public health (mainly contagious diseases and slaughtering), and nuisance legislation.[30] British legislators wrestled with the idea of cruelty to animals, as evidenced by the many attempts to amend and refine the offense during the nineteenth century.[31] Efforts

---

28. Charles Darwin, for example, proposed to use the BAAS report as a model for the regulation of vivisection in his letter to T. H. Huxley, January 14, 1875, Darwin Correspondence Project, "Letter no. 9817," https://www.darwinproject.ac.uk/letter/DCP-LETT-9817.xml.

29. Edward Emanuel Klein, Sir Michael Foster, and Sir T. Lauder (Thomas Lauder) Brunton, *Handbook for the Physiological Laboratory*, ed. J. Burdon-Sanderson (Philadelphia: Lindsay and Blakiston, 1873).

30. For example, An Act to Amend the Law for Regulating Places Kept for Slaughtering Horses, 1844, 7&8 Vict. c. 87; Sea Birds Preservation Act, 1869, 32 & 33 Vict. c. 17; The Contagious Diseases (Animals) Act, 1869, 32&33 Vict. c. 70.

31. For an extensive survey of the laws, see Brian Harrison, "Animals and the State in Nineteenth-Century England," *English Historical Review* 88, no. 349 (October 1973): 786–820. Harrison emphasizes the active role of the Royal Society for the Prevention of Cruelty to Animals in the legislation. Harrison, "Animals and the State," 789.

to outlaw cruelty to animals were already underway at the turn of the century but passed Parliament only in 1822 with the first Act to Prevent the Cruel and Improper Treatment of Cattle.[32] Also known as Martin's Act after its avid advocate MP Richard Martin, the first anticruelty law made it an offense to "wantonly and cruelly beat, abuse or ill treat any horse, mare, gelding, mule, ass, ox, cow, heifer, steer, sheep and other cattle."[33]

An 1833 law made bear-baiting and cockfighting in the London metropolis illegal, since at those places "idle and disorderly persons commonly assemble; to the interruption of good order and the danger of public peace."[34] This provision, which was initially part of a broader public order scheme that included restrictions on coffee houses and public horn-blowing, was repealed in 1835 and reintegrated into an amended version of Martin's Act. The Act to Consolidate and Amend the Several Laws Relating to the Cruel and Improper Treatment of Animals, and the Mischiefs Arising from the Driving of Cattle, and to make Other Provisions in Regard Thereto added the word "torture" to the list of illegal behaviors and extended the list of protected animals to dogs "or any other cattle or domestic animal."[35] The 1835 Act also specified new obligations, including the feeding and watering of confined animals and the licensing of slaughterhouses. A new version of the anticruelty law was introduced in 1849 as An Act for the More Effectual Prevention of Cruelty to Animals. It eliminated the word "wantonly" from the previous legislation, added over-driving to the list of prohibited behaviors, and added cats to the list of protected animals.[36] After six years the law was amended again to include all domestic animals.[37]

---

32. An Act to Prevent the Cruel and Improper Treatment of Cattle, 1822, 3 Geo. IV. C. 71. About Richard Martin, see Kathryn Shevelow, *For the Love of Animals: The Rise of the Animal Protection Movement* (New York: Holt Paperbacks, 2009), 182–200.

33. Martin's Act, § 1.

34. An Act for the More Effectual Administration of Justice in the Office of a Justice of the Peace in the Several Police Offices established in the Metropolis, and for the More Effectual Prevention of Depredation on the River Thames and its Vicinity, for Three Years, 1833, 3 W. 4, c. 19.

35. An Act to Consolidate and Amend the Several Laws Relating to the Cruel and Improper Treatment of Animals, and the Mischiefs Arising from the Driving of Cattle, and to make Other Provisions in Regard Thereto, 1835, 5&6 W. IV., C. 59.

36. An Act for the More Effectual Prevention of Cruelty to Animals, 1849, 12&13 Vict. c. 92. Justice Day explained in *Lewis v. Fermor* that the omission of "wantonly" was insignificant. Lewis v. Fermor (1887) 18 Q.B.D. 53.

37. An Act to Amend an Act of the Twelfth and Thirteenth Years of Her Present Majesty for the More Effectual Prevention of Cruelty to Animals, 1854, 17&18 Vict. c. 60. The 1849 Act was repealed and replaced by the Protection of Animals Act, 1911, 1 & 2 Geo 5., c. 27. The same year of the Vivisection Act, Parliament legislated The Drugging of Animals Act, 1876, 39 Vict. c. 13. The Act prohibited the poisoning of "horses and other animals by disqualified persons," which meant a person who was not the owner.

British courts developed the "unnecessary suffering" doctrine in a succession of precedents before it was implanted in cruelty to animals legislation. The Society for the Prevention of Cruelty to Animals (later the Royal Society for the Prevention of Cruelty to Animals, RSPCA) initiated most indictments under the prevention of cruelty laws. Established in 1824 following Martin's Act, the society undertook to educate the public and to enforce the law by employing a private police force.[38] Justice Whitman set the basic formulation of the "unnecessary suffering" doctrine in the 1867 case *Budge v. Parson*. The case evolved around cockfighting, and the court decided that a person who set his cock to fight against an injured cock committed an offense under the 1849 Act. Justice Whitman commented that "the cruelty intended by the statute is the unnecessary abuse of any animal."[39] From then on, the necessity test was a principal element in the cruelty offense.

In 1874 came the first opportunity to implement criminal law and the doctrine of "unnecessary suffering" in the case of vivisection. The RSPCA prosecuted the French physician Eugene Magnan and four other medical Englishmen for violating the Cruelty to Animals Act (1849) during a demonstration at a meeting of the British Medical Association. But a substantial discussion failed to evolve. The magistrates dismissed the case, known as the Norwich trial, unconvinced that the defendant was the one to conduct the controversial experiments. The court remarked that the case was proper to prosecute and therefore did not impose any costs on the RSCPA.[40] The trial was formative for future efforts to legislate against vivisection. For antivivisectionists, the dismissal of the case unraveled the shortcomings of the existing legal framework of cruelty to animals when dealing with animal experimentation.[41]

---

38. Harrison, "Animals and the State," 794. The society was first named Society for the Prevention of Cruelty to Animals, and endorsed by the Queen in 1825. Ritvo, *Animal Estate*, 129. For the role of the private organizations in the implementation of animal laws, see Jessica Wang, "Dogs and the Making of the American State: Voluntary Association, State Power, and the Politics of Animal Control in New York City, 1850–1920," *Journal of American History* 98, no. 4 (March 2012): 998–1024.

39. Budge v. Parsons (1863) 3 B. & S. 382, 146.

40. "Prosecution At Norwich. Experiments On Animals," *British Medical Journal* 2, no. 728 (1874): 754.

41. Susan Hamilton, introduction to *Animal Welfare & Anti-Vivisection 1870–1910: Frances Power Cobbe* (New York: Routledge, 2004), xxii. British judges continued to develop the "unnecessary suffering" doctrine in the criminal law in the years following the Vivisection Act, mirroring the same utilitarian logic and confronting similar challenges with calculating animal pain. The next influential opinion was given in an 1877 case of comb cutting, Murphy v. Manning (1877) 2 Exch. D. 307. Justice Kelly used the necessity test: "The question is, is there any purpose or reason which can legalise or justify an act of such extreme barbarity." Justice Cleasby stressed again that cruelty was an "unnecessary abuse." The Queen's Bench Division of the High Court elaborated the doctrine of unnecessary suffering in the 1889 landmark case of Ford v. Wiley (1899) 23 Q.B.D. 203. The court was asked to

The Legislative History of the Act

Among the various explanations for the appeal of the anticruelty movement, historians agree that the protection of animals was embraced as a virtue by middle and upper-class patrons, and often functioned in the hands of humanitarians as an "instrument of marginalization" of the lower classes.[42] However, vivisection was an oddity among the practices attacked by animal lovers. Those who conducted it were "from those layers of society from whom moral guidance was traditionally sought."[43] Historians traced some of the critics' discomfort with vivisection to their distrust of the medical profession, resulting from dreadful experiences in the treatment of women and the poor—an unease that sometimes converged with resistance to vaccination.[44] This mistrust joined a more profound attack in the name of religious benevolence on modern values of progress and interest-driven ideology.[45] National identity and anti-French sentiments played into public criticism of vivisection as well: British scientists adopted experimental physiology only in the mid-nineteenth century when it was already flourishing in Continental sciences. The antivivisection campaign was powered

decide whether the sawing of oxen's horns in order to facilitate their collective confinement was cruel according to the law. Justice J. Hawkins in *Ford v. Wiley* proclaimed that the relevant circumstances for the cruelty offense were "the amount of pain caused, the intensity and duration of the suffering, and the object sought to be attained." He elaborated on the proportionality test, according to which "the beneficial or useful end sought to be attained must be reasonably proportionate to the extent of the suffering caused, and in no case can substantial suffering be inflicted, unless necessity for its infliction can reasonably be said to exist" (220). By explaining that pain should be inflicted reasonably and in a proportional volume to a desirable and legitimate object, the decision in *Ford v. Wiley* embedded the calculability of suffering in animal welfare law, reinforcing the logic that directed the implementation of the Vivisection Act.

42. Ritvo, *Animal Estate*, 130. Class dimension and economic calculus are also discussed in Thomas, *Man and the Natural World*, 185–88. Kathleen Kete describes an analogous process of class formation and the design of Parisian bourgeois identity in the French context. Kathleen Kete, *The Beast in the Boudoir: Petkeeping in Nineteenth-Century Paris* (Berkeley: University of California Press, 1995), 21.

43. Hilda Kean, "The 'Smooth Cool Men of Science': The Feminist and Socialist Response to Vivisection," *History Workshop Journal*, no. 40 (October 1995): 20. See also Ritvo, *Animal Estate*, 157.

44. The mistrust was largely connected to body-trafficking networks; see Elizabeth T. Hurren, *Dying for Victorian Medicine: English Anatomy and Its Trade in the Dead Poor, c.1834–1929* (Houndmills: Palgrave Macmillan, 2014), 5. The distrust toward the medical establishment was intertwined with women's struggle for autonomy. Coral Lansbury, *The Old Brown Dog: Women, Workers, and Vivisection in Edwardian England* (Madison: University of Wisconsin Press, 1985): 183. For connections between human experimentation and vivisection in the US context, see Susan E. Lederer, *Subjected to Science: Human Experimentation in America before the Second World War* (Baltimore: Johns Hopkins University Press, 1997), 101–25. For the connections between antivivisection and antivaccination, see Nadja Durbach, *Bodily Matters: The Anti-Vaccination Movement in England, 1853–1907* (Durham: Duke University Press Books, 2004), 113–49.

45. Ritvo, *Animal Estate*, 164.

by a grassroots movement, led by women and clergy. The movement also enjoyed the Crown's support, and in 1875 Queen Victoria wrote to surgeon Joseph Lister urging him to denounce the practice of vivisection.[46]

Two competing private bills concerning vivisection were circulating in mid-1875, one supported by the scientists' lobby, presented by Lyon Playfair, and the other by advocates of the abolition of vivisection, presented by Lord Henniker. The bills stirred up public debate and led Prime Minister Benjamin Disraeli to initiate an inquiry into the practice of animal experimentation. On May 24, 1875, Home Secretary Richard Cross announced the Royal Commission on the Practice of Subjecting Live Animals to Experiments for Scientific Purposes.[47] Aiming to "consider whether any and what measures ought to be adopted in respect to that practice," the commission interviewed fifty-three witnesses, including "many eminent physicians and surgeons and physiologists of great reputation."[48] Scientists played at least two roles on the witness stand. They were in the position of educators guiding the commission through the complexities of animal bodies and providing information about the vivisection practice—its scope, purpose, and benefits.[49] But the scientists testifying in front of the commission were also, or at least they felt as if they were, potential suspects. Many assured the commission that the British medical students were reluctant to hurt animals. The Royal Commission could then proudly state that "at the present time a general sentiment of humanity on this subject appears to pervade all classes in this country."[50]

---

46. "Let Us Now Praise Famous Men," *British Medical Journal* 1, no. 3457 (1927): 677. The historians Richard French, Harriet Ritvo, and Susan Hamilton provide a thorough examination of the individuals, institutions, agendas, and beliefs that fed into the antivivisection agitation. About the transformation in British attitudes toward animals, see Keith Thomas, *Man and the Natural World: Changing Attitudes in England, 1500–1800* (New York: Pantheon Books, 1983), 143–50; Ritvo, *Animal Estate*, 126; Mike Radford, *Animal Welfare Law in Britain: Regulation and Responsibility* (Oxford: Oxford University Press, 2001), 19–27. For vivisection and British imperialism, see Peter Hobbins, *Venomous Encounters: Snakes, Vivisection and Scientific Medicine in Colonial Australia* (Manchester: Manchester University Press, 2017); Pratik Chakrabarti, "Beasts of Burden: Animals and Laboratory Research in Colonial India," *History of Science* 48, no. 2 (June 1, 2010): 125–51.

47. About Cross's selection process for the Royal Commission, see Richard French, *Antivivisection and Medical Science in Victorian Society* (Princeton, NJ: Princeton University Press, 1975), 92–96.

48. Report of the Royal Commission on the Practice of Subjecting Live Animals to Experiments for Scientific Purposes; with Minutes of Evidence and Appendix (London: Eyre & W. Spottiswoode, 1876), vii (hereinafter: "Royal Commission Report," or "Royal Commission").

49. For an analysis of some of the testimonies, see Rob Boddice, "Species of Compassion: Aesthetics, Anaesthetics, and Pain in the Physiological Laboratory," *19: Interdisciplinary Studies in the Long Nineteenth Century*, no. 15 (June 2012), 1–22.

50. Royal Commission Report, x. Richard French's monograph from 1975, the most comprehensive study of the Act, provides a detailed account of the struggles behind the statute. French, *Antivivisection and Medical Science*, 112–43.

In January 1876, the commission submitted its report, which concluded with a call for legislation.[51] The report recommended placing experiments on animals under the control of the home secretary, who would have the power to grant licenses. The licenses, which would be mandatory for experimenters, would include conditions "to ensure that suffering should never be inflicted in any case in which it could be avoided, and should be reduced to a minimum where it could not be altogether avoided." The Royal Commission advised that "in the administration of the system generally, the responsible minister would of course be guided by the opinion of advisers of competent knowledge and experience." The home secretary "must have the most complete power of efficient inspection," and places where experiments were performed should be registered for that cause. With these and additional measures, "the progress of medical knowledge may be made compatible with the just requirements of humanity."[52]

The Royal Commission's report was endorsed by the Victorian antivivisection movement, as well as by the scientific community at large. Cobbe, who founded in December 1875 the Victoria Street Society for the Protection of Animals from Vivisection, called on the government to accept the report's recommendation.[53] The British Association for the Advancement of Science also endorsed the report. The Scottish physiologist John Grey M'Kendrick stated in his address to the Department of Anatomy and Physiology in its 1876 annual meeting that "it seems to me that the appointment of a Royal Commission to investigate the facts of the case was the best thing that could have been done by the government . . . and gave in a report which, while it recommended legislation, is generally in favour of physiologists."[54] Although experimenters were initially alarmed by the legislative attempts to monitor their practice, they ended

51. French, *Antivivisection and Medical Science*, 96.

52. Royal Commission Report, xx, xxi.

53. Cobbe, Frances Power. *The Life of Frances Power Cobbe, by Herself* (Boston: Houghton and Mifflin, 1894), 2:588; French, *Antivivisection and Medical Science*, 112. Cobbe was an outstanding figure in the antivivisection movement; see Sally Mitchell, *Frances Power Cobbe: Victorian Feminist, Journalist, Reformer* (Charlottesville: University of Virginia Press, 2004). See also Moira Ferguson, "Frances Power Cobbe: Antivivisection, Feminism, and Nationhood," in *Animal Advocacy and Englishwomen, 1780–1900: Patriots, Nation, and Empire* (Ann Arbor: University of Michigan Press, 1998), 105–24; Susan Hamilton, "Reading and the Popular Critique of Science in the Victorian Anti-Vivisection Press: Frances Power Cobbe's Writing for the Victoria Street Society," *Victorian Review* 36, no. 2 (October 2010): 66–79. For an edited collection of Cobbe's writings, see Susan Hamilton, *Animal Welfare & Anti-Vivisection 1870–1910: Nineteenth Century Woman's Mission*, vol. 1 (London: Routledge, 2004).

54. Report of the Forty-Sixth Meeting of the British Association for the Advancement of Science Held at Glasgow in September 1876 (London, 1877), 128. The reactions to the Royal Commission Report were not united in either of the parties; see French, *Antivivisection and Medical Science*, 110–11.

up gaining more than they had expected. As historians of medicine show, following the antivivisection assault and subsequent regulation, the scientific lobby was consolidated, and medical research went through professionalization and institutionalization. The scientific lobby was consolidated largely as a response to the regulation attempts.[55]

Even though the report was generally well-received, major disagreements surfaced once it came to actual legislation. Following the commission's recommendation for legislation, Conservative Lord Carnarvon, the colonial secretary, drafted a bill primarily based on a proposal by the RSPCA. This organization was somewhat critical but mostly tolerant of animal experimentation.[56] Carnarvon's Bill added controversial provisions to the Royal Commission's recommendations, such as a restriction of lawful experiments to medical objectives, and the exclusion of cats and dogs from vivisection.[57] Debates went on until the very last moments before the legislation was passed; physiologists disapproved even of the title of the Act. In a copy of the bill kept by the physician Lauder Brunton, a thin black ink line crossed out the word "cruelty" and replaced it with "experiments," suggesting that the Act should be cited as "Experiments on Animals Act" rather than "Cruelty to Animals Act."[58]

The final version of the Act—see appendix II for the full text—imposed the following restrictions on the performance of any experiment calculated to inflict pain upon a living animal: the experiment had to be performed with a view to the advancement of physiological knowledge, or of knowledge which would be useful for saving or prolonging life or alleviating suffering; a person had to hold a license to experiment on animals, granted by the secretary of state (home secretary); the animal had to be under the influence of anesthetics, while curare would not be considered anesthetic in this context; the animal had to be killed before recovery from anesthesia, if the pain was likely to continue; and the experiment should not be performed to attain a manual skill.

A list of exemptions modified these restrictions, and the Home Office designated a special certificate for each: experiments could be performed without anesthesia under a certificate affirming that insensibility could not be produced without necessarily frustrating the objects of such experiments (certificate A);

---

55. Rob Boddice, *Humane Professions: The Defence of Experimental Medicine, 1876–1914* (Cambridge: Cambridge University Press, 2021).

56. French, *Antivivisection and Medical Science*, 108, 115; RSCPA, Précis of Evidence to Be Given before the Royal Commission on Vivisection, November 11, 1907, HO 144/4. All references to HO (Home Office) department code in the following notes are from the British National Archives.

57. French, *Antivivisection and Medical Science*, 115.

58. Bill intituled An Act to Prevent Cruel Experiments on Animals, Wellcome Archives (WA), SA/RDS/A3.

the obligation to kill the animal before it recovered from anesthetic could be waived under a certificate affirming that the killing of the animal would necessarily frustrate the object of the experiment, and provided that the animal be killed as soon as the object was attained (certificate B); experiments might be performed as illustration for lectures under a certificate affirming that the experiment was absolutely necessary (certificate C); experiments might be performed not directly for the advancement of a new discovery of physiological knowledge, but for the testing of such knowledge, under a certificate affirming that the testing was absolutely necessary for the effectual advancement of knowledge (certificate D). The Act prescribed a certificate for experimenting on dogs and cats without anesthesia, affirming that the object of the experiment would necessarily be frustrated unless it was performed on an animal similar to these animals and that no other animal was available for the experiment, (certificate E) and a certificate for painful experiments on horses, asses, or mules (certificate F).[59] The Home Office later designated an additional certificate—unspecified in the Act—for allowing the recovery of dogs and cats after anesthesia (certificate EE). The Act outlawed the use of the substance curare as anesthesia and the exhibition of experiments to the public. The penalty for violating the Act was not to exceed £50.

Notwithstanding that the Act did not apply to invertebrate animals (Section 22), it was species-inclusive compared to other anticruelty legislation. The implementation of the Act was the responsibility of the home secretary, who was provided with extensive leeway in designing his department's policy. He was allowed to require the registration of places used for experimentation, to license any person whom he found qualified, to limit the duration of the license, and to revoke licenses. The home secretary was also provided with the power to annex conditions for licenses and to ask licensees for reports on the results of their experiments.[60] Moreover, a prosecution under the Act against a licensed person needed the assent of the home secretary.[61]

---

59. The British Medical Association opposed the requirement for a special certificate in addition to the license. The Parliamentary Bills Committee of the British Medical Association recommended omitting from the Bill "all the clauses which require special certificates in addition to licenses for performing painful operations and further special certificates in respect to dogs, cats, etc." Parliamentary Bills Committee of the British Medical Association, "The Cruelty to Animals Bill," n.d., WA, SA/RDS/A/3. About certificate C and the teaching of physiology, see E. M. Tansey, "'The Queen Has Been Dreadfully Shocked': Aspects of Teaching Experimental Physiology Using Animals in Britain, 1876–1986," *American Journal of Physiology* 274, no. 6 (June 2, 1998): S21–S25.

60. Vivisection Act, Sections 7–9.

61. Vivisection Act, Section 21. In Ireland, the administration of the Act was under the Chief Secretary to the Lord Lieutenant. See also Section 20.

The Curare Provision

The events leading to the composition of the curare provision in the Vivisection Act are revealing in terms of the melding of law and scientific knowledge. Curare was a paralyzing poison derived from South American plants. For centuries it fascinated European explorers with its deadly powers. Today, curare is understood to operate as a muscle relaxant through neuromuscular blocking: a chemical compound, *d*-Tubocurarine, that disconnects the motor nerves from the muscles. The result is a temporary paralysis, while the sensory neurons remain unaffected. However, eighteenth- and nineteenth-century explorers and physiologists interpreted the curarized animal's paralysis and inability to express pain as numbness. The difficulty in identifying the pain of others, the debates within the scientific community about the effects of the poisonous substance on sentience, and the confrontation between scientists and legislators about the use of curare, foreshadowed the issues that would trouble the Act's implementation.[62]

A set of experiments done in the mid-1850s convinced the French physiologist Claude Bernard that curare does not render the animal insentient.[63] Meanwhile, vivisectors had discovered that when administered in the right quantities, curare made invasive procedures much more manageable. Due to its relaxing effects on the muscles, curare had become a popular device to make animals motionless during experimentation. Adding tubocurarine allowed surgeons to minimize the quantities of anesthetics needed to make a body easy to handle, and thus reduced the risks involved in deep anesthesia for human and nonhuman bodies.[64]

Bernard's interpretation of curare's action on the nerves made waves on the other side of the canal. Most important for animal welfare advocates, Bernard's experiments showed that animals under the influence of curare were deprived of the ability to show signs of pain. Some physiologists testified in the Royal Commission's meetings that curare worked as an anesthetic and spared the animal the sensation of pain. For example, Lauder Brunton, then a lecturer in materia medica and pharmacology at Saint Bartholomew's Hospital and a fellow of the Royal Society, opined that Bernard was mistaken. Brunton claimed that he succeeded in demonstrating "that wourali certainly paralyses the sensory nerves as well as the motor."[65] Prominent medical figures joined this interpretation.

---

62. Shira Shmuely, "Curare: The Poisoned Arrow That Entered the Laboratory and Sparked a Moral Debate," *Social History of Medicine* 33, no. 3 (August 1, 2020): 881–97.

63. Shmuely, "Curare," 446–47.

64. Keith Sykes and John Bunker, *Anaesthesia and the Practice of Medicine: Historical Perspectives* (London: Royal Society of Medicine Press, 2007), 106.

65. Royal Commission Report, 285. For an illuminating analysis of the testimonies about curare, see Daniel Hoffman, "Fatal Attractions: Curare-Based Arrow Poisons, from Medical Innovation to

The histologist and bacteriologist Edward Emanuel Klein, a contributor to the vivisection manual mentioned above, contended that curare diminished suffering in frogs, referring the commission to experiments made by Moritz Schiff. For Klein, pain and movement were interconnected: "We take it generally that when we pinch the frog and it does not move it does not feel."[66]

Francis Sibson, a physician at St. Mary's Hospital, member of the Senate of the University of London, and fellow of the Royal Society, presented an original idea about pain under curare. He claimed that even if curare did not render the body insentient, it still lessened the experience of pain, since "the animal would be perfectly at rest; there would be no struggling; the incisions would be made with great ease." Ascribing to curare somewhat meditative powers, Sibson maintained, "the animal mind would be withdrawn from what I would call the domain of attention and where there is no attention there is no sensation, no sensitiveness, no pain."[67]

Others expressed an ambivalent opinion yet were inclined to believe curare had a certain effect on the nervous system. The renowned physiologist William Sharpey claimed that there was not enough data to determine how curare operates. But the Commission Chairman Viscount Cardwell pushed Sharpey on that point, raising the question of public interest. The chairman asked, "If the public have any rights at all in this matter to have their feeling respected, those rights are not regarded when the experiments performed under worari poison are held out as experiments performed under anaesthetic?" Sharpey had to admit he thought "that is not a sufficient answer to the public."[68] Flipping the roles between animals and vivisectors, Commissioner William Edward Forster suggested that curare alleviated the *experimenter's* distress by preventing the animal from crying.[69]

The claim that curare had anesthetic qualities was highly contested; some witnesses argued—as did Claude Bernard—that curare did not affect the sensory nerves. The curarized animal feels pain, they argued, but cannot express it. George Hoggan, self-described as the first to bring up the question of curare abuse

Lethal Injection" (PhD diss., University of California, Berkeley, 2009), 123–86; Murrie, "*Death-in-Life*," 264–66.

66. Royal Commission Report, 188, 189.

67. Royal Commission Report, 238.

68. Royal Commission Report, 21, 22.

69. Royal Commission Report, 153. About pain in the Royal Commission, see Rob Boddice, "Species of Compassion: Aesthetics, Anaesthetics, and Pain in the Physiological Laboratory," *19: Interdisciplinary Studies in the Long Nineteenth Century*, 15, 2012, 3–6.

in England, expressed full confidence in Bernard's theory.[70] Similarly, Arthur de Noe Walker, a private practitioner in London, was convinced that curare was not an anesthetic and that "its use for that purpose should be forbidden."[71]

The Royal Commission acknowledged in its report that the poison was "very convenient to an operator, since it paralyses the motor nerves, and keeps the animal quiet." But its usefulness for physiological research was overshadowed by the findings of Bernard, "perhaps the highest authority on such a subject." As the commission explained, "An animal may be suffering exquisite torture and yet (so far as we yet know) the worari poison may, by its effect upon the motor nerves, prevent demonstration of any feeling." The epistemic doubt also existed regarding the converse event, when "an animal may make every demonstration of suffering while the real sensation is destroyed." These uncertainties in identifying pain were not restricted to animals. Even with human patients under chloroform, or those suffering from an injury to the spine, "it sometimes happens that all outward manifestation of pain is exhibited, when the patient afterwards disclaims having experienced any sensation of it."[72]

It should be noted that the commission was not unanimous in its findings. From the sessions' proceedings, it is clear that Thomas Henry Huxley advocated a different view. Huxley criticized Bernard, arguing the latter had shown at most that curare disturbed the reflex action. Huxley contended that Bernard had "jumped to the conclusion" that curare did not affect sentience, in a publication that was "perhaps more worth to the pen of Victor Hugo than of that of a staid physiologist."[73] Nevertheless, the commission recommended that until the dispute had been settled, curare would not be considered an anesthetic by law.

The commission's recommendation was integrated into the vivisection bill, presented by the Earl of Carnarvon in May 1876. Clause four of the bill read, "The Substance Known as urari or curare shall not for the purposes of this Act be deemed to be an anesthetic." Meaning that a painful experiment performed with curare and no other anesthetic would be a violation of the law, unless under a special certificate that allows for experimentation without anesthesia.[74] The Medical Council opposed the proposed restrictions on the use of curare in physiological practice. Responding to the Royal Commission Report and the

---

70. Royal Commission Report, 204–5; see also Susan Hamilton, *Animal Welfare & Anti-Vivisection 1870–1910: Frances Power Cobbe* (New York: Routledge, 2004), xx.

71. Royal Commission Report, 91.

72. Royal Commission Report.

73. Royal Commission Report, 239.

74. *Cruelty to Animals Bill* (HL) (Bill 250, 1876).

subsequent bill, the council carried a motion contending that "in the present state of knowledge as to the properties of woorari, it is inexpedient for Parliament to declare, as in this Bill, that this drug shall not, for the purpose of the Act, be deemed anaesthetic." The council suggested the alternative, according to which: "Neither urari nor any other substance should be deemed to be an anaesthetic for the purpose of the Act until it has been proved to be so."[75]

From the physiologists' point of view, they were fighting against unwelcome interference in their practice, as much as for their authority in matters of science—a realm to which they believed the question of pain and the use of curare belonged. An editorial pronounced this view at the Manchester Guardian, stating, "The Bill itself starts by deciding for us a controverted point in regard to the properties of the substance called urari or wourali, or curari or curare, whose very name appears to be in dispute." Interpreting curare was not a matter for lawmakers, "This assumed power to determine a scientific controversy by Act of Parliament is nothing by the side of the delegated power which it is proposed to give to certain authorities to determine all manner of controversies at their own discretion."[76] Despite the resistance of the scientific lobby, the Royal Commission's recommendation was accepted, and the skepticism about curare's abilities to make animals insentient was written into the law. The Vivisection Act clarified that "the substance known as urari or curare shall not for the purposes of this Act be deemed to be an anaesthetic."[77]

The Review Process at the Home Office

The Act passed in August 1876, and by November printed copies of certification forms were available to purchase at Messrs Chronicle Medical Publishers. Certificate forms were sold for a halfpenny in three locations in London, Dublin, and Edinburgh, and at times were provided upon request. License forms opened with the statement "I beg to apply for a licence to perform experiments under the above-mentioned Act," followed by a blank space in which the applicant had to insert his name and profession, "the nature, general description, number, and purpose of the experiments intended, the place where it is proposed to perform them, the time for which the license is required, and any other circumstance that

75. "General Council of Medical Education and Registration," *BMJ*, 1876, 1, 689, 700, 696.
76. Editorial, *The Manchester Guardian*, May 20, 1876, 7.
77. Cruelty to Animals Act, 1876, 39&40 Vict. c. 77, Section 4.

may be deemed necessary." At the bottom, there was a space for the authorized persons' signatures under the sentence, "we recommend that the above application be granted." A copy of Section 11 of the Act, detailing who was authorized to sign the application, was printed on the back of the form. Licenses were valid for no more than twelve months.

A similar template was used for application forms for the various certificates, adapted for each exception. Thus, for example, in certificate B that allowed for the recovery of an animal after an experiment, the authorized persons signed the statement: "We hereby certify that, in our opinion, the killing of the animals on which any such experiment is performed before it recovers from the influence of the anaesthetic administered to it would necessarily frustrate the object of such experiment." Certificate D, "for the purpose of testing the former discoveries," contained a two-column table—one "description of proposed experiments" and one "description of former discoveries for the purpose of testing which the proposed experiment are to be made." All certificates provided limited space for describing the proposed experiment, allowing for no more than a few sentences. While the home secretary granted the licenses, two statutory authorities signed off the certificates: one signature by a president of a scientific society, and one by a professor of physiology, medicine, anatomy medical jurisprudence, materia medica, or surgery. The home secretary had the power to disallow or suspend the certificates given by the above persons.

The home secretary had ways to control the certificates, even as scientific authorities granted them. He could disallow them but also, as explained in a later official report, could exercise indirect control since a certificate was valid only with a license. The home secretary could annex conditions to the license that would affect the certificate, and hereby "exert almost as complete a control over the working of a certificate, once given, as he can over a licence." In the first two decades, the practice was to limit certificates by time or number of experiments. In 1900, the time limitation was relaxed, and certificates were given for five years. Limitation on the number of animals was particularly enforced in experiments under certificate B (which allowed for the recovery of the animal from anesthesia,) or in other certificates that involved the use of cats, dogs, and monkeys.

A review process for the applications was established at the Home Office. Each license application was examined by one of the inspectors, who provided the home secretary with a recommendation. At times the inspectors received questions and requests from physiologists. In such cases they either responded directly or referred the inquiry to the undersecretaries. Legal advisers and undersecretaries from the Home Office also addressed inquiries and other correspondence

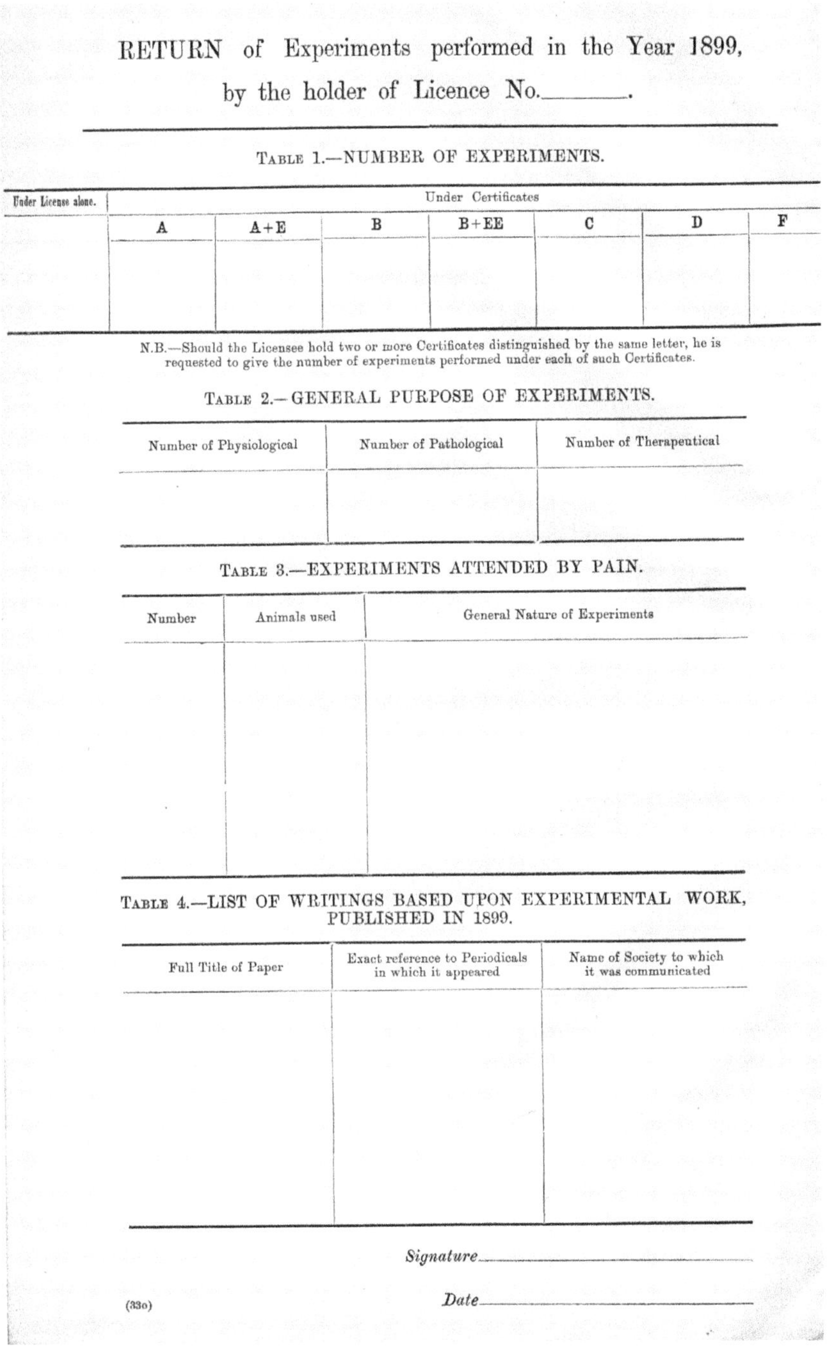

RETURN of Experiments performed in the Year 1899, by the holder of Licence No._____.

TABLE 1.—NUMBER OF EXPERIMENTS.

Under Licence alone.	Under Certificates						
	A	A+E	B	B+EE	C	D	F

N.B.—Should the Licensee hold two or more Certificates distinguished by the same letter, he is requested to give the number of experiments performed under each of such Certificates.

TABLE 2.— GENERAL PURPOSE OF EXPERIMENTS.

Number of Physiological	Number of Pathological	Number of Therapeutical

TABLE 3.—EXPERIMENTS ATTENDED BY PAIN.

Number	Animals used	General Nature of Experiments

TABLE 4.—LIST OF WRITINGS BASED UPON EXPERIMENTAL WORK, PUBLISHED IN 1899.

Full Title of Paper	Exact reference to Periodicals in which it appeared	Name of Society to which it was communicated

Signature_____

(33o)

Date_____

FIGURE 2 / A form for returns of experiments to be filled by a licensed experimenter, 1899, TNA, HO 144/451/B30824.

regarding the provisions of the Act from various sources, including physiologists, medical organizations, and antivivisection societies.[78]

In order to manage the system of licenses and certificates, the Home Office made an additional requirement from the experimenters to create a small-scaled cataloging and recording system in their laboratories. In March 1877, the Home Office provided licensees with forms to keep records of all the experiments they performed under their license and certificates, with instructions to forward the records to the inspector upon request (the "blue forms").[79] In 1902, the practice changed so that experimenters were no longer asked to use the Home Office forms to keep records of their experiments, if they kept the records in a book in the laboratory.[80]

In addition to these records, the Home Office required licensees to submit yearly reports on all their experiments to the inspectors. The terminology of the forms changed. In 1899, for example, the legal assistant undersecretary Henry Cunynghame and the inspector George Thane delivered circular letters reminding licensees to transmit to the latter the "Annual Returns of Experiments," without which their licenses would not be renewed. The annexed form contained four tables. First, a table of the number of experiments under each certificate. Second, a table of "General Purpose of Experiments," divided into physiological, pathological, and therapeutic types. Third, a table of "Experiments Attended by Pain," which included the number, animals used, and a rubric for describing the "General Nature of Experiments." The fourth table was a "List of Writings Based upon Experimental Work" published the same year.[81] The Home Office also required licensees to send copies of any published descriptions of experiments including published lectures or private circulation. These materials were,

---

78. Home Secretary H. A. Bruce initiated the senior post of a legal adviser in 1869. Jill Pellew, "Law and Order: Expertise and the Victorian Home Office," in *Government and Expertise: Specialists, Administrators and Professionals, 1860–1919*, ed. Roy MacLeod (Cambridge: Cambridge University Press, 2003), 65.

79. Letter template, circa March 1877, HO 156/1, 70; Inspector James Russell explained that "white forms" were "filled up simply with the number of experiments, and whether they are pathological, physiological, or whatever they may be." The "blue form" was "the large form in which they enter the date, and the animal and the certificate, with or without anaesthesia, and what have done to it, and the final result." Royal Commission on Vivisection, *Appendix to First Report of the Commissioners: Minute of Evidence, October to December 1906* (London: Wyman and Sons, 1907), 27, http://wellcomelibrary.org/item/b28038496 (hereinafter cited as Royal Commission on Vivisection, *Appendix to First Report*).

80. Memorandum of Evidence to be Given by Mr. W. P. Byrne, C.B., A Principal Clerk in the Office of H.M. Secretary of State for the Home Department, HO 114/4; Royal Commission on Vivisection, *Final Report of the Royal Commission on Vivisection* (London: Wyman and Sons, 1912), 415 (hereinafter cited as Royal Commission on Vivisection, *Final Report*).

81. Letter template by Thane, Return of Experiments Performed in the Year 1899, December 16, 1899, HO 144/451/B30824.

according to the Home Office, "carefully scrutinized by the Inspectors with a view to the satisfaction of the Secretary of State that no procedures causing pain have been adopted without due authority."[82]

Although the Act allowed the home secretary to direct a licensee to submit the results of experiments, it was rarely done. William Byrne, assistant undersecretary to the home secretary, explained:

> The matter has received anxious consideration at the Home Office. It has been felt that the Secretary of State would be placed in a position of grave responsibility and much difficulty if he were to undertake to formulate and act on a Departmental opinion as to whether the scientific results of a research—possibly of an obscure character and conducted by an expert of unique qualifications—were sufficiently established or promising to justify the pursuit of the investigation. And it would be unreasonable to expect that his Inspectors, however learned and accomplished, could be at home in every branch of the most modern research in physiology, pathology, pharmacology, and the rest.[83]

Soon after the passing of the Act, the inspectors used the data they collected from the licensees to compose their Annual Reports to the Parliament. Anthony Mundella, MP for Sheffield, initiated the House of Commons' order of returns in February 1877. Mundella asked to list the name of licensees; registered places; the certificates permitting experiments as illustrations of lectures to students; the certificates permitting experiments on cats, dogs, horses, mules, or asses; the certificates permitting performing experiments without anesthetics, and the number of such experiments in which curare had been employed; and the scientific authorities who had in each case granted such certificates.[84] Home Secretary Cross replied he "had no objection to any Return that might be asked for to call into account the action of the Secretary of State in the administration either of this or of any other Act," and agreed to produce returns "in a somewhat altered form." Cross preferred to omit the names of licensees from the returns, and instead to report only the number of experimenters holding each certificate. Mundella accepted Cross's modification of his request, and the first annual returns were published on May 21, 1878 without the names of licensees.[85]

---

82. Second Royal Commission, *Final Report*, 9.
83. Second Royal Commission, *Final Report*.
84. 232 Parl. Deb. (3rd ser.) (1877) 634.
85. Cruelty to Animals: Return of Licenses Granted under the Act 39 & 40 Vict. c. 77, to Amend the Law Relating to Cruelty to Animals, Parliamentary Papers, 1878, (193).

However, in December 1878, the assistant undersecretary (and later a permanent undersecretary) Adolphus Liddell instructed inspector George Busk to prepare "a most *careful report*," including the number of experiments, experimenters, and their names. Liddell required Busk to "most clearly distinguish between painful and painless experiments."[86] The second return was published with the names of licensees on March 27, 1879, together with a short explanation by inspector Busk about the data and his evaluation of painless and painful experiments.[87] In the following years, the Home Office published yearly returns combined with explanatory reports by the inspectors.

The Vivisection Act was tailored to settle the conflicting demands of a critical public and developing biomedical research through parliamentary craftwork. At the same time, it was a legal embodiment of utilitarian medicine, reflecting the aspiration to nullify surplus pain while keeping the animal functioning to produce scientific knowledge. The actualization of these ideals over the reality of living bodies confronted experimenters and civil servants alike with fundamental questions about the definition of pain, its expressiveness, and its malleability.

86. Liddell to Busk, December 19, 1878, HO 156/1, 255.

87. Experiments on Animals: Copy of any Report from the Inspectors showing the Number of Experiments Performed on Living Animals during the Year 1878, under Licences Granted under the Act 39 & 40 Vict. c. 77, Distinguishing Painless from Painful Experiments, Parliamentary Papers, 1878–79, (127).

THE RIGHT FORMS FOR THE JOB /
Anesthesia, Brain Research, and Certificate E

On a wintery Saturday in December 1847, a few men gathered around a lame horse in East Lothian, east of Edinburgh. The horse's owner, Mr. Reid, was there, and so were two medical doctors. They soaked a flannel cloth in two ounces of chloroform and placed it in a tin case. They then covered the horse's nose with it, and in three and a half minutes saw the animal collapse. The *Illustrated London News* reported the "interesting and most successful experiment with this beneficent agent," which enabled a veterinary surgeon to operate on the horse's forefeet and cut through its nerves "without a quiver."[1] The horse was sound and back on his legs again in twenty-five minutes. Chloroform, which was synthesized in 1831, joined other anesthetics that were making their first steps in medical treatment. Experimenters noted the anesthetic qualities of nitrous oxide ("laughing gas") already in 1800, but surgeons took advantage of them only in the mid-century. On October 16, 1846, a year before Mr. Reid chloroformed his horse, the dentist William T. G. Morton administered ether during a surgery at the Massachusetts General Hospital in arguably the best-known public demonstration of anesthesia. The news spread widely, and a couple of months later, the surgeon Robert Liston employed ether during an amputation in London, opening a new era for surgeons, suffering bodies, and the regulation of pain.

The technology of anesthesia, which was developed by scientists for medical treatment and incorporated into physiological research as a restraining device, turned into an ethical instrument with its obligatory use under subsection 3(3) of the Act: "The animal must during the whole of the experiment be under the influence of some anesthetic of sufficient power to prevent the animal feeling pain." Many scientists, state administrators, and mainstream animal welfare advocates were united in their belief that anesthesia could be the solution to the problem of pain in laboratories. The Act's embrace of anesthesia drew from the assumption that animal pain could be contained through medical and

---

1. "Application of Chloroform to Animals," *Illustrated London News*, December 11, 1847, 381.

legal means. Even radical antivivisectionists such as Frances Power Cobbe, who referred to anesthesia as a "curse to the vivisectible animals," admitted at times that theoretically, a wholly anesthetized animal—which she doubted could be attained—would disarm her resistance.[2] A coalesced law-and-science operation thus aimed to create, using anesthesia, the ultimate experimental subject: a painless animal.[3]

For this reason, when the anesthetizing of an animal was inapplicable for one reason or another, unrest was felt at the offices of the home secretary. Such moments of disillusionment multiplied in the last two decades of the nineteenth century as brain research gained momentum. Disputes between the Home Office and physiologists over issuing licenses and certifying neurological experiments manifested the legal and conceptual problems with the unanesthetized animal. While debating how to classify procedures on the brain that necessitated keeping the animals alive, law and science co-constructed concepts of pain and experiment. Only "experiments calculated to give pain" came under the Act. But what if the vivisector contended that his experiment was over, while the civil servant had reasons to believe it was still going on? The definition of experiment projected back to the question of pain—was it the same creature as before the operation, was its pain a concern for the law, were experimenters still obliged to treat a recovered animal as instructed by the Act? Brain research was polemic not only because it was hard to classify under the Act, but also because it often involved dogs, whose pain evoked special sentiments in Victorians.

Anesthesia: Killing the Pain, Keeping the Animal

Anesthesia not only debunked conceptions about the inevitability of pain but also contributed to the transformation in the perception of pain from a valuable element in healing to a physiological obstacle that could and should be eliminated. In 1853, Queen Victoria invited physician Dr. John Snow (1813–58) to assist with the birth of her fourth child, Prince Leopold. Snow administered chloroform to

---

2. Victorian Street Society for the Protection of Animals from Vivisection united with the International Association for the Total Suppression of Vivisection, *Anaesthetic and Vivisection*, n.d., 1–2; Frances Power Cobbe and Benjamin Bryan, *Vivisection in America: I. How It Is Taught II. How It Is Practiced* (London: Swan Sonnenschein, 1890), 5.

3. Tarquin Holmes and Carrie Friese analyze the role of the anesthetized animal as a boundary object, in Tarquin Holmes and Carrie Friese, "Making the Anaesthetised Animal into a Boundary Object: An Analysis of the 1875 Royal Commission on Vivisection," *History and Philosophy of the Life Sciences* 42, no. 4 (October 14, 2020): 50.

the Queen, keeping her under the influence of anesthesia for nearly an hour.[4] Upon her death in January 1901, the *British Medical Journal* commemorated "the courage of the Queen, and her confidence in her physicians" expressed on the occasion. When praising the Queen's courage, the journal alluded not only to her willingness to try chloroform but also to her disregard of religious objections to the use of anesthetic during childbirth.

Interest in the amelioration of human pain was not new, but previous attempts to ease the pain of sufferers were overshadowed by the long-held belief that pain played an indispensable part in the healing process. Enduring pain had theological significance and was considered essential for spiritual redemption, particularly in pivotal life experiences such as childbirth and on the deathbed.[5] Victorians treated pain with a special urgency, and its relief gained increasing importance in medical care.[6] Scholars debate about how best to explain the growing nineteenth-century sensitivity to pain. The historian Stephanie Snow supports Roy Porter's observation that during the nineteenth century, the "very pain threshold of society" was lowered.[7] However, the sociologist and legal scholar Shai Lavi contests the attempt to explain the sensitivity to pain as an instance of Norbert Elias's "civilizing process," according to which Western societies have become less comfortable with extreme experiences. Instead, Lavi argues that there was a cultural rather than individual sensitivity to pain and a fundamental change in the place of pain in the world; the problem of pain "was not the intensity but intelligibility, not sensitivity but sensibility."[8]

---

4. "Death of Queen Victoria," *British Medical Journal* 1, no. 2091 (1901): 234. See also Stephanie J. Snow, *Blessed Days of Anaesthesia: How Anaesthetics Changed the World* (Oxford: Oxford University Press, 2009), 88–91. An anthology by Rachel Ablow, *Pain, Subjectivity, and the Social* (Princeton, NJ: Princeton University Press, 2017), examines pain in Victorian literature.

5. Guerrini, *The Courtiers' Anatomists*, 79; Snow, *Blessed Days of Anaesthesia*, xii. About the significance of pain at the deathbed, see Shai Lavi, *The Modern Art of Dying*, 65; Donald Caton, *What a Blessing She Had Chloroform: The Medical and Social Response to the Pain of Childbirth from 1800 to the Present* (New Haven: Yale University Press, 1999), 960. Snow gives as an example the evangelists' encouragement of using opiates on the deathbed, to provide the dying person with the conditions to repent. Stephanie J. Snow, *Operations without Pain: The Practice and Science of Anaesthesia in Victorian Britain* (Basingstoke: Palgrave Macmillan, 2005), 19; Roselyne Rey, *The History of Pain* (Cambridge: Harvard University Press, 1998), 91.

6. Snow, *Operations without Pain*, 4.

7. Snow, *Operations without Pain*, 19; Roy Porter, *Flesh in the Age of Reason: The Modern Foundations of Body and Soul*, rev. ed. (New York: W. W. Norton, 2004), 403.

8. Shai Lavi, *The Modern Art of Dying*, 68. Studies that focus on the political and cultural history of pain include Javier Moscoso, *Pain: A Cultural History* (Houndmills: Palgrave Macmillan, 2012). The anthropologist Adriana Petryna elucidates how pain and suffering were created "within scientific/social orders." Adriana Petryna, *Life Exposed: Biological Citizens after Chernobyl* (Princeton: Princeton University Press, 2013), 12. Keith Wailoo argues that the attitudes toward pain changed with the transformations in the "model sufferer." Wailoo also shows that "the courts (far more than

Anesthesia also marked a new era for those who treated suffering bodies. Victor Robinson explained in his 1946 *Victory over Pain* that "to the surgeon, pain was the barrier he was forced to penetrate before his terrible instrument became the healing knife. The pain of the patient often denied access to his body."[9] Ether and chloroform removed the pain barrier and paved the way for entering the body's most hidden organs and delicate tissues. While anesthesia helped surgeons mechanically by constraining the otherwise struggling body, it also alleviated some of the stress surgeons experienced when operating on grieving and struggling patients.[10]

This double effect of anesthesia was also true for performers of vivisection. Together with an elaborate catalog of restraining devices, anesthesia eased the access to animals' bodies by helping vivisectors to hold the animal still, as testified by manuals of physiological research, which advised students to use it for this purpose.[11] Dror explains that "pain distorted normal physiology and hindered successful replication. Physiological knowledge demanded pain-free animals."[12] At the same time, as noted by Guerrini, anesthesia "freed the conscience of some scientists," increasing the number of invasive experiments.[13] Anesthetics, therefore, joined other techniques that suppressed or managed the feelings of experimenters toward their animal subjects.[14] But the analogy between the role of anesthesia in human medical care and animal experimentation should not be overemphasized. For human patients, anesthesia provided a temporary stage

physicians) that have decided, time and time again, how to measure distress and how to define the right to relief." Keith Wailoo, *Pain: A Political History* (Baltimore: Johns Hopkins University Press, 2014), 7, 207.

9. Victor Robinson, *Victory over Pain, a History of Anesthesia* (New York: Schuman, 1946), 3. Historiographically, Robinson's work belongs to a group of studies that emphasized the achievements of anesthesia, such as the works of MacQuitty and Sykes and Bunker cited above.

10. Snow, *Operations without Pain*, 24.

11. Edward Klein and John Burdon-Sanderson, eds., *Handbook for the Physiological Laboratory* (Philadelphia: Lindsay & Blakiston, 1873). For an analysis of the role of the *Handbook* in British physiology and the vivisection debates, see Stewart Richards, "Drawing the Life-Blood of Physiology: Vivisection and the Physiologists' Dilemma, 1870–1900," *Annals of Science* 43, no. 1 (January 1986): 27.

12. Otniel E. Dror, "The Affect of Experiment: The Turn to Emotions in Anglo-American Physiology, 1900–1940," *Isis* 90, no. 2 (June 1999): 210.

13. Guerrini, *Experimenting with Humans and Animals*, 81. See also Stewart Richards, "Anaesthetics, Ethics and Aesthetics: Vivisection in the Late Nineteenth-Century British Laboratory," in *The Laboratory Revolution in Medicine*, ed. Andrew Cunningham and Perry Williams (Cambridge: Cambridge University Press, 2002), 142–69.

14. About managing the feelings of Victorian physiologists in laboratories, see Paul White, "Sympathy under the Knife: Experimentation and Emotion in Late Victorian Medicine," in *Medicine, Emotion and Disease, 1700–1950* (New York: Palgrave Macmillan, 2006), 101. Rob Boddice suggests thinking about nineteenth-century physiologists as practicing compassion toward the whole community. Boddice, "Species of Compassion," V.

in their path to healing, whereas for animals, it was often a step toward their destruction.

Alongside the secularization of suffering in nineteenth-century Britain and its vanishing spiritual significance, science found a role for pain. In the chapter "Comparison of the Mental Powers of Man and the Lower Animals" in *The Descent of Man* (1871), Charles Darwin argued, "the lower animals, like man, manifestly feel pleasure and pain, happiness and misery . . . The fact that the lower animals are excited by the same emotions as ourselves is so well established, that it will not be necessary to weary the reader by many details." After discussing emotions of terror, suspicion, courage, and timidity, he used a scene of vivisection to exemplify dogs' love: "Every one has heard of the dog suffering under vivisection, who licked the hand of the operator; this man, unless the operation was fully justified by an increase of our knowledge, or unless he had a heart of stone, must have felt remorse to the last hour of his life."[15] The dog's licking gesture expressed emotions, and this message triggered, or should have triggered, according to Darwin, care and sympathy.

Darwin dedicated the subsequent *The Expression of the Emotions in Man and Animals* (1872) to further analyzing the performativity of emotions. Pain has a role for the sufferer by assisting her in avoiding risky situations: "Great pain urges all animals, and has urged them during endless generations, to make the most violent and diversified efforts to escape from the cause of suffering." Pain triggers somatic responses of forceful muscle activity in the chest and vocal muscles, "and loud, harsh screams or cries will be uttered." Next to the practical aim of freeing oneself from a painful situation, the bodily movements have a role in communicating the pain and in soliciting assistance: "But the advantage derived from outcries has here probably come into play in an important manner; for the young of most animals, when in distress or danger, call loudly to their parents for aid, as do the members of the same community for mutual aid."[16] When publishing *The Expression*, Darwin's prominent position in British science was beyond dispute. His message had doubtlessly resonated among British educated people: Humans and other animals share pain; pain might not redeem the soul, but it has a role in protecting the individual creature; pain is highly expressive and communicable.

Darwin's attentiveness to animal pain made the antivivisection leader Francis Power Cobbe believe that the famed scientist would be her ally in the demand

---

15. Charles Darwin, *The Descent of Man and Selection in Relation to Sex* (London: John Murray, Albermile Street, 1871), 1:39, 40.

16. Charles Darwin, *The Expression* (London, 1872), 72.

to outlaw vivisection—she and Darwin were corresponding on professional and friendly terms for years; he even sent her a copy of *The Expression* shortly after publication.[17] However, Darwin's ethical calculus showed him otherwise. To Cobbe's dismay, Darwin refused to sign a memorial demanding the Royal Society for the Prevention of Cruelty to Animals (RSPCA) to act against vivisection. In a letter to Cobbe, dated January 14, 1875, Darwin expressed his sentiments toward suffering creatures, "That any experiment should be tried without the use of anaesthetics, when they can be used, is atrocious." However, he declined to join Cobbe's campaign, explaining, "I believe that Physiology will ultimately lead to incalculable benefits, and it can progress only by experiments on living animals. Any stringent law would stop all progress in this country which I should deeply regret."[18]

A few months later, on November 3, 1875, Darwin restated his views before the Royal Commission on Vivisection. Again, he distinguished between experiments with and without anesthetics. Regarding experiments on insensible animals, he said, "It is unintelligible to me how anybody could object to such experiments. I can understand a Hindoo, who would object to an animal being slaughtered for food, disapproving of such experiments, but it is absolutely unintelligible to me on what ground the objection is made in this country." Nevertheless, when committee member Viscount Cardwell asked about his view on "painful experiment without anaesthetics, when the same experiment could be made with anaesthetics," Darwin replied that it "deserves detestation and abhorrence."[19]

The Royal Commission embraced the view that pain is the only moral problem with vivisection and that anesthesia could relieve the ethical concerns and, ultimately, the critique over animal experimentation. Even the RSPCA expressed no objection to experiments under anesthetics, and as its representative later clarified, "the attitude taken up by the Society, almost since its foundation, had been that it deprecates all experiments on animals which cause pain, but as regard experiments which cause no pain there is no ground for interference

---

17. F. P. Cobbe to Darwin, November 26, 1872. Darwin Correspondence Project, "Letter no. 8649," https://www.darwinproject.ac.uk/letter/?docId=letters/DCP-LETT-8649.xml&query=8649.

18. Charles Robert Darwin to Frances Power Cobbe, January 14, 1875, Darwin Correspondence Project, "Letter no. 9814F," https://www.darwinproject.ac.uk/letter/DCP-LETT-9814F.xml. For more about Darwin's position in the vivisection debates, see Chris Danta, "The Metaphysical Cut: Darwin and Stevenson on Vivisection," *Victorian Review* 36, no. 2 (October 2010): 51–65; Rod Preece, "Darwinism, Christianity, and the Great Vivisection Debate," *Journal of the History of Ideas* 64, no. 3 (2003): 399–419; David Allan Feller, "Dog Fight: Darwin as Animal Advocate in the Antivivisection Controversy of 1875," *Studies in History and Philosophy of Science Part C: Studies in History and Philosophy of Biological and Biomedical Sciences* 40, no. 4 (December 2009): 265–71.

19. Royal Commission 1875, 234.

THE RIGHT FORMS FOR THE JOB

by the Society, because the question of cruelty does not arise."[20] Member of the Royal Commission and future Home Office inspector John Eric Erichsen referred to the introduction of anesthetics as a "new era" for animal experimentation.[21] The commission commented in its report that vivisection "has been or at least ought to have been, relieved of the greater part of its difficulty by the discovery of anaesthetics."[22]

Anesthesia, however, failed to deliver its promise for science with no suffering. Although early doubts about anesthesia's efficacy were relaxed with the accumulating testimonies of human patients, its use also had troublesome side effects.[23] Some even criticized the Act for its mandatory anesthetic provision, contending that it was a paradoxical requirement that made animals worse off. Moreover, for a whole class of experiments, anesthesia was simply inapplicable. Increasingly during the last two decades of the nineteenth century, operations on the nervous system required the monitoring of alert and active animals. The cracks that threatened to break the illusion of a painless animal were quickly filled with forms, conditions, and legal opinions.

Gerald Yeo and the Certificate to Experiment on Dogs and Cats

The physiologist Gerald Francis Yeo (1845–1909) earned his MB (Bachelor of Medicine, Bachelor of Surgery) and MCh (Master of Surgery) in 1867 at Trinity College, Dublin. He practiced and taught physiology in Dublin, later traveling to universities in Paris, Berlin, and Vienna. Yeo pursued an MD at Dublin in 1871, and in the following year became a member of the Royal College of Physicians and a member of the Royal College of Surgeons in Ireland. In 1874, on the backdrop of an evolving antivivisection agitation, Yeo was appointed professor of physiology at King's College, London. On March 31, 1876, Yeo joined eighteen other British scientists at the home of the physiologist John Burdon Sanderson (1828–1905) to discuss a response to the Royal Commission Report

---

20. RSPCA, "Précis of Evidence to be Given before the Royal Commission on Vivisection," November 11, 1907, HO 144/4.

21. Royal Commission Report, 3.

22. Royal Commission Report, xi. Guerrini claims that the introduction of anesthesia released scientists from previous constraints, and led to an increase in invasive experiments. It also forced antivivisectionists to re-examine their oppositional approach to experiments. Guerrini, *Experimenting with Humans and Animals*, 81.

23. Snow, *Operations without Pain*, 4.

and the expected legislation. They founded the Physiological Society, of which Yeo was the first secretary.[24] After practicing as an assistant surgeon in King's College Hospital, he was appointed a fellow at the Royal College of Surgeons. Yeo resigned from his clinical appointment in 1880 to devote his time to physiological research. Yeo was elected a fellow of the Royal Society in 1889 and was a prolific experimenter on varied topics such as bile, heart muscles, and skeletal muscles, although, as the *British Medical Journal* remarked in his obituary, "not a voluminous writer of original papers."[25]

Yeo's longtime acquaintance, the physiologist Edward Sharpey-Schäfer (1850–1935), described him as "pleasant and friendly but impetuous, and unyielding in argument, steadfastly declining to be 'convinced against his will.'"[26] Yeo's interaction with the Home Office was characterized by repeated disagreements, to which deeper tensions between the Home Office and the scientific lobby were channeled. The disagreements between Yeo and the Home Office focused on the interpretation of one section of the Act, but they brought to the surface substantial disparities in the understanding of animals and the nature of experiments.

Yeo's first conflict with the Home Office emerged soon after the Act came into effect. In December 1877, Yeo applied for certificate B—a form that would allow an animal to recover from anesthesia—for a planned experiment on dogs. Yeo described the proposed experiment as creating a fistula of the small intestine and noted in the form that the animal would be under chloroform during the entire operation. Robert McDonnell, the president of the Royal College of Surgeons, and John Curnow, a professor of anatomy at King's College, signed the standard statement: "We hereby certify that, in our opinion, the killing of the animal on which any such experiment is performed before it recovers from the influence of the anaesthetic administered to it would necessarily frustrate the object of such experiment."[27]

Assistant undersecretary Godfrey Lushington (1832–1907) examined the request. Lushington was a son of a Whig-liberal family, with a dominant presence at the bar and in the public service. He joined the Home Office as a legal adviser in October 1869 and was promoted to legal assistant undersecretary

24. Edward Sharpey-Schafer, "History of the Physiological Society during Its First Fifty Years, 1876–1926," *Journal of Physiology* 64, no. 3 (1927): 5.

25. "Gerald Francis Yeo, M.D., F.R.S., Emeritus Professor of Physiology, King's College, London," *British Medical Journal* 1, no. 2523 (1909): 1158.

26. Sharpey-Schafer, "History of the Physiological Society during Its First Fifty Years," 32.

27. Gerald Yeo, Form of Certificate B23, December 7, 1877, HO 144/17/44209F.

in 1876, a period that the historian Jill Pellew describes as the "beginning of twenty years in which the Home Office departmental view was particularly legalistic."[28] Before joining the Home Office, he taught at the Working Men's College in London, wrote for the Positivist Society, and was involved during the 1860s in the trade union movement. "One of the earliest English converts to the religion of humanity," in Pellew's words, Lushington was a member of the Social Science Association's committee on trade societies and strikes, the Jamaica Committee, and the Emancipation Society.[29] In sixteen years of service as an under-secretary and ten as a permanent under-secretary, Lushington enjoyed an influential position in the department, and his opinions and recommendations were highly appreciated by the home secretaries he worked for. He was knighted in 1892.[30]

Lushington did not approve Yeo's application for certificate B and asked him to supplement it with certificate E, as required for experiments on dogs without anesthesia, for each animal he planned to experiment upon.[31] Certificate E corresponded to Section 5 in the Act, according to which "an experiment calculated to give pain shall not be performed without anaesthetics on a dog or cat," except on such a certificate stating that "the object of the experiment will be necessarily frustrated unless it is performed on an animal similar in constitution and habits to a cat or dog, and no other animal is available for such experiment."

Given the legal history of anti-cruelty laws, it is not surprising that the Act designated certain species for additional protection. Martin's Act of 1822 had been restricted to a wide array of horses and cattle, as well as sheep. One of the initial objectives of the Society for the Prevention of Cruelty to Animals was to broaden the scope of protection to other species.[32] Dogs were added by the Cruelty to Animals Act (1835), which listed "horse, mare, gelding, bull, ox, cow, heifer, steer, calf, mule, ass, sheep, lamb, dog, or any other cattle or domestic

---

28. Jill Pellew, "Law and Order: Expertise and the Victorian Home Office," in *Government and Expertise: Specialists, Administrators and Professionals, 1860–1919*, ed. Roy MacLeod (Cambridge: Cambridge University Press, 1988), 67.

29. Jill Pellew, "Sir Godfrey Lushington," in *DNB*, http://www.oxforddnb.com.libproxy.mit.edu/view/article/38727; "Obituary: Sir Godfrey Lushington," *The Working Men's College Journal: Conducted by Members of the Working Men's College* (London, 1907).

30. Pellew, "Law and Order," 67. Lushington's twin brother, lawyer Vernon Lushington, was an acquaintance of Charles Darwin. It seems like their relationship did not involve discussing the vivisection controversy, although see Darwin to J. S. Burdon Sanderson, April 11, 1875, Darwin Correspondence Project, "Letter no. 9923."

31. Lushington to Yeo, December 11, 1877, HO 156/1, 159.

32. Shevelow, *For the Love of Animals*, 270.

animal."[33] In 1839 the use of draft dogs was abolished, and in the Cruelty to Animals Act (1849), cats were mentioned explicitly for the first time.[34]

It was a time when pet-keeping became widespread and emotionally charged.[35] Middle-class Victorian pet fanciers established a pet hierarchy with their emphasis on breeding and pedigreed animals, facilitating the establishment of new canine and feline taxonomies. Conversely, farmers and hunters tended to think about stray dogs and cats as vermin. They detested affectionate pet relations and had an interest in dogs and cats only so far as they could fulfill their role in the field.[36] These distinctions within the categories, however, were not reflected in humane legislation at that stage, including the Vivisection Act, which referred to all dogs and cats regardless of their breed or origin.

The Royal Commission of 1875 did not recommend restricting experiments on cats, dogs, or any other kind of animal. The requirement for certificate E was initiated rather as a response to committee member Richard Holt Hutton's minority report, in which he opined that the law should prohibit experimentation on dogs and cats. Hutton, a journalist and a theologian, praised the relationship of trust and servitude between humans and cats and dogs, arguing that these special relations rendered it immoral to turn cats and dogs into instruments for scientific research; he claimed that there was "something in the nature of treachery as well as insensibility to their suffering" in subjecting them to experiments. Hutton also warned that allowing such experiments would sanction the stealing of these companions causing "a great distress to the owners," and stressed that domestication had made cats and dogs more sensitive to pain than other species, a state he referred to as "hyperaesthesia": "No other class of animals otherwise convenient for experimentation contains so many creatures of high intelligence, and therefore probably of high sensibility, as dogs and cats."[37] In other words,

---

33. Act to Consolidate and Amend the Several Laws Relating to the Cruel and Improper Treatment of Animals, and the Mischiefs Arising from the Driving of Cattle, and to Make Other Provisions in Regard Thereto, 1835, 5&6 W. IV., c. 59.

34. An Act for Further Improving the Police in and Near the Metropolis, 1839, 2&3 Vict. c. 47. In its original version, the ban was in the context of urban nuisance, but in the Cruelty to Animals Act (1854) it was restated as a protection measure.

35. Harriet Ritvo, *The Platypus and the Mermaid: And Other Figments of the Classifying Imagination* (Cambridge: Harvard University Press, 1998), 86. Victorian pet-keeping influenced similar US trends after World War I; see Jones, *Valuing Animals*, 116; Katherine C. Grier, *Pets in America: A History* (Chapel Hill: University of North Carolina Press, 2006). For pet keeping as the creation of common group consciousness in France, see Kete, *The Beast in the Boudoir*; for pet keeping in the colonial context, see Shmuely, Wallace's Baby Orangutan, JHB 2020.

36. Ritvo, *Animal Estate*, 95, 115.

37. Richard Holt Hutton, January 8, 1876, Royal Commission Report, xxii; Susan Hamilton, introduction to *Animal Welfare & Anti-Vivisection 1870–1910*, xxii; Guerrini, *Experimenting with Humans and Animals*, 81; White, *The Experimental Animal*, 68.

Hutton argued not only that humans bear ethical obligations toward cats and dogs constituted by their shared history, but also that cats and dogs are more prone to suffer pain by virtue of domestication.

The first draft of the bill incorporated Hutton's recommendation and banned the use of cats and dogs for experimentation. The proposed ban was one of the most divisive elements in the bill. A group of "teachers of physiology in England, Scotland, and Ireland," among whom were William Sharpey, Michael Foster, and Gerald Yeo, composed a "Memorandum of Facts and Considerations Relating to the Cruelty to Animals Bill." In the memorandum, which was presented to Home Secretary Richard Cross, they argued that exempting cats and dogs from experimentation "would entirely prevent some of the most important investigations in physiology and in medicine."[38]

Cats and dogs were precious for medical research because they were "the only flesh-eating, warm-blooded animals which are available for observation in England. Agreeing in these respects with man, these animals (and especially the dog) resemble him in their digestion, assimilation, secretion, nutrition and the chemical processes of life generally." The nervous system and "processes of fever and its relation to bodily temperature and circulation" were additional points of resemblance. Dogs were also necessary for the comparative study of disease, which required the use of various species. Zoonotic diseases, such as rabies, and external parasitic diseases, required research on dogs and cats to fight human maladies. The memorandum further explained that while rodents were prone to tuberculosis, dogs had more cancer than many other animals. The memorandum concluded the section on dogs and cats by reassuring its readers that "these animals can be rendered completely and continuously insensible to pain by the use of suitable anaesthetics."[39] Along the same lines, the General Medical Council composed a memorial claiming there was no "sufficient reason for specially exempting dogs and cats from experiments."[40] The physiologist John Burdon Sanderson, a central figure in the scientists' lobby opposing the bill, suggested a compromise: replacing the ban on research on dogs and cats with a requirement for a special certificate. Responding to the continual pressure from medical

---

38. Memorandum of Facts and Considerations Relating to the Practice of Scientific Experiments on Living Animals, Commonly Called Vivisection, n.d., WA, SA/RDS/A/3.

39. Memorandum of Facts and Considerations Relating to the Practice of Scientific Experiments on Living Animals, Commonly Called Vivisection, undated, WA, SA/RDS/A/3,4; French, *Antivivisection and Medical Science*, 125.

40. Memorial from the General Medical Council to Her Majesty's Government Respecting the Bill Intitled "An Act to Prevent Cruel Experiments on Animals," n.d., Royal College of Physicians (RCP), 2021/7, 5.

organizations, and despite the opposition of antivivisectionists, the bill was amended to allow experiments on dogs and cats under a special certificate.[41]

Although they proposed the special dogs and cats certificate, physiologists tried to avoid its use, mainly out of a preference not to have an official record associating them with this controversial practice. Based on this, Yeo refused Lushington's request to submit certificate E, explaining that he had "no desire to perform any experiment on dogs and cats without anaesthetics and therefore I do not require that certificate."[42] Yeo planned to anesthetize the dogs and then let them recover. At first, the Home Office did not accept Yeo's logic and insisted he should submit the certificate "because although the experiment itself is performed under anesthetics, the dog is not being destroyed after the effect of the anaesthetic has passed off."[43]

Yeo wrote back, stressing that he could not apply for certificate E since he did "not want or intend to perform experiment upon any kind of animal *without anaesthetics*." Yeo doubted that the persons qualified to sign off on certificates would endorse such a "paradoxical statement," since in his proposed experiment, the animal would be under the influence of anesthetics "so long as the proceedings are calculated to give pain."[44] In Yeo's view, the experiment period was limited to the time of the operation on the body, while the period following the recovery of the animal from anesthesia was an observation rather than painful experimentation and thus required no special certificate.

A consultation took place at the Home Office. In a memo directed to Home Secretary Cross, Lushington admitted that there was "much to be said for Dr. Yeo's contention." The question was "whether the after effects of the experiment viz the pain which the animal suffers or the serious injury under which the animal labors after the anaesthetic has gone off, are to be deemed part of the experiment calculated to give pain, although nothing more in the nature of experiment has to be performed."[45] Lushington acknowledged that certificate E for experimenting without anesthesia on dogs and cats suggested that the experiment was done without anesthesia at all. He also comprehended the possible refusal of leading physicians to sign off on such a form, as well as Yeo's reluctance to apply for one.

---

41. French, *Antivivisection and Medical Science*, 121.
42. Yeo to Lushington, December 14, 1877, HO 144/17/44209F.
43. Memo by Godfrey Lushington, December 14, 1877, HO 144/17/44209F; Liddell to Yeo, December 30, 1877, HO 156/1, 162.
44. Yeo to the Secretary of State, December 27, 1877, HO 144/17/44209F.
45. Memo by Lushington, n.d., HO 144/17/44209F.

Lushington deliberated that a solution to Yeo's objection could be replacing the wording of the standard form, "to perform on dogs and cats the experiments . . . without anaesthetic," with the statement that the certificate allows researchers to "abstain from killing a dog or cat after the same has recovered from the influence of the anaesthetic under which the experiment described below shall have been performed." But that would mean departing from the exact language of the Act—something that the Home Office wished to avoid. Setting aside the fundamental questions that Lushington raised, the home secretary decided to allow Yeo to perform his experiment on dogs and cats without certificate E. The decision was based on the assumption that "the animal will not really suffer pain as the after effects of the experiment."[46]

During the following years, Yeo developed an interest in brain research. In May 1881, Yeo published a "Note on the Application of the Antiseptic Method of Dressing to Craniocerebral Surgery" in the *British Medical Journal*. Assisted and advised by his King's College colleague, neurologist David Ferrier, Yeo performed the experiment on monkeys "being the animals which, in septic reception, as well as their behavior under chloroform and operation, most resemble man." Another possible reason was that at that time, experimenting on monkeys did not require special certificates from the Home Office. Yeo added in the publication that "as far as possible, the animals were treated in the same way as human subjects"—a comment that can be read both as describing his scientific method and as a statement reassuring his more critical readers that the experiment was done with care.[47] Yeo would use almost the same phrasing in his next certificate application.

In late October 1881, Yeo applied for a certificate B (allowing an animal to recover from anesthesia) for a brain experiment. The aim was "to test the efficacy of the antiseptic method of dressing wounds of the cranium and the membranes of the brain," with the clarification that "the animals being treated in all possible respects as human subjects would be after similar operation or injury."[48] This time inspector George Busk (see chapter 4) was the one to process Yeo's request. Busk initially recommended against granting the certificate because Yeo had previously held certificates of the same kind and should have indicated "what advancement or additional information can be obtained" from conducting the

46. Memo by Lushington, n.d., HO 144/17/44209F; Lushington informed Yeo about Cross' decision in Lushington to Yeo, January 3, 1878, HO 156/1, 172.

47. Gerald F. Yeo, "Note on the Application of the Antiseptic Method of Dressing to Craniocerebral Surgery," *British Medical Journal* 1, no. 1063 (1881): 763.

48. Gerald Yeo, Form of Certificate B, October 25, 1881, HO 144/17/44209F.

experiment. Ten days after his initial recommendation, however, Busk changed his mind, explaining that he had been mistaken when suggesting that Yeo had already performed similar experiments.[49] Yeo's certificate was finally approved, and he was drawn into a much larger drama.[50]

The Trial of Dr. Ferrier

Yeo's arrival at King's College started off productive, doing collaborative research with neurologist David Ferrier (1843–1928). Ferrier had earned an MA in 1863 in classics and philosophy at Aberdeen University. He spent a year studying psychology at Heidelberg and completed an MB in medicine in 1868 at the University of Edinburgh. In 1871, Ferrier joined King's College as a demonstrator of physiology, and a year later was appointed a professor of forensic medicine. In 1873, the Scottish physician and psychiatrist James Crichton-Browne (1840–1938), now famous for his neuropsychiatric photography, invited Ferrier to conduct brain research in the facility he directed, West Riding Lunatic Asylum at Wakefield.[51] During his residence, Ferrier explored the possibility of triggering epileptic seizures from the cortex using mild alternating currents to stimulate different areas in the exposed brains of rabbits, cats, and dogs. It was a theatrical scene, as described by the historian Stanley Finger: "His animals walked, grabbed, scratched, blinked, and even flexed their digits."[52]

In the following years, Ferrier shifted his focus to macaque monkeys to refine his brain mapping.[53] The sociologist Susan Leigh Star describes Ferrier's notes from those years as "often written in an obviously hasty, shaky hand, trying to record minute-by-minute events in the laboratory: the pages are spattered with blood stains."[54] Some of the electrical irritation was employed on

---

49. Memo, November 17, 1881, HO 144/17/44209F.

50. As indicated by the Annual Report for 1882. A note next to Yeo's certificate B reads "this Certificate is no longer in force." Experiments on Living Animals: Copy of Report from Inspectors Showing the Number of Experiments Performed on Living Animals during the Year 1882, under Licences Granted under the Act 39 & 40 Vict. c. 77, Distinguishing Painless Experiments from Painful Experiments, Parliamentary Papers, 1883 (176).

51. Images of the residents of West Riding Lunatic Asylum were used, among others, by Darwin in *Expression of Emotions*.

52. Stanley Finger, *Minds behind the Brain: A History of the Pioneers and Their Discoveries* (Oxford: Oxford University Press, 2000), 163.

53. David Ferrier, "Experiments on the Brain of Monkeys.—No. I," *Proceedings of the Royal Society of London* 23 (1874): 409–30.

54. Susan Leigh Star, "Scientific Work and Uncertainty," *Social Studies of Science* 15, no. 3 (August 1, 1985): 403.

conscious monkeys. For example, Ferrier reported about one monkey, "which was allowed to remain quite conscious during stimulation, an experiment was made as to vision by holding before it a teaspoonful of milk, which it was eager to seize. In its attempt this point was stimulated, with the effect of causing confusion of vision and some difficulty in reaching the milk." In another episode, "The application of the electrodes to the ganglia on the left side . . . caused the animal to utter various barking, howling, or screaming sounds of an incongruous character."[55]

The prevalent theory in neuroscience up until the 1870s was that the cerebral cortex was unexcitable and that brain functions were diffused in its various parts. The studies conducted by Ferrier demonstrated that lesions of the brain could not only simulate controlled movement but also produce sensory and mental reactions and transform the animal's character and disposition. In his 1875 Croonian lecture, Ferrier explicated: "I selected the most active, lively, and intelligent animals which I could obtain. To one seeing the animals after the removal of their frontal lobes little effect might be perceptible, and beyond some dulness and inactivity they might seem fairly up to the average of monkey intelligence. They seemed to me, after having studied their character carefully before and after the operation, to have undergone a great change."[56] Victorian novelists were inspired by Ferrier's scientific persona and were especially intrigued by his alteration of experimental creatures' character and mentality through brain manipulation. Wilkie Collins modeled Dr. Nathan Benjulia after Ferrier in *Heart and Science: A Story of the Present Time* (1883), and literary scholars detected the neurologist's influence also in H. G. Wells's *The Island of Dr. Moreau* (1896) and Bram Stoker's *Dracula* (1897).[57] Ferrier's experimental work and its clinical implications, particularly the brain localization maps he produced, dominated

---

55. Ferrier, "Experiments on the Brain of Monkeys," 425, 429.

56. David Ferrier, "The Croonian Lecture: Experiments on the Brain of Monkeys (Second Series)," *Philosophical Transactions of the Royal Society of London* 165 (1875): 433–88, 440.

57. Jessica Straley, "Love and Vivisection: Wilkie Collins's Experiment in Hearth and Science," *Nineteenth-Century Literature* 65, no. 3 (December 1, 2010): 354; A. Stiles, "Cerebral Automatism, the Brain, and the Soul in Bram Stoker's Dracula," *Journal of the History of the Neurosciences* 15 (2006), 131–52; Anne Stiles, *Popular Fiction and Brain Science in the Late Nineteenth Century* (Cambridge: Cambridge University Press, 2011); Laura Otis, "Howled out the of Country: Wilkie Collins and H.G. Wells Retry David Ferrier," in *Neurology and Literature, 1860–1920*, ed. Anne Stiles (Houndmills: Palgrave Macmillan, 2007). About Ferrier's symbolism in Victorian literature and antivivisection imagination, see also Michael A. Finn and James F. Stark, "Medical Science and the Cruelty to Animals Act 1876: A Re-Examination of Anti-Vivisectionism in Provincial Britain," *Studies in History and Philosophy of Science Part C: Studies in History and Philosophy of Biological and Biomedical Sciences* 49 (February 2015): 12–23.

cerebral physiology during the 1870s and remained in use well into the twentieth century.[58]

On October 27, 1875, while a professor of forensic medicine at King's College and an assistant physician at King's College Hospital, Ferrier testified before the Royal Commission. He expressed his opposition to the regulation of vivisection, claiming that "any legislation that would retard physiological research would be a discredit to this country."[59] He also objected to inspection "of any kind." Nevertheless, Ferrier testified that he "certainly always endeavored to prevent unnecessary suffering in any animal upon which I did experiment." The commission confronted Ferrier with an accusation made by John Colam, the RSPCA's secretary, according to which Ferrier's experiments on monkeys stimulated laughter in his audience. Colam contended that Ferrier's demonstrations were "amusement offered to the audience" that betrayed his indifference to animal pain as much as the vulgar character of the event, a claim that was often directed to organized animal fights. Ferrier rejected the accusation: "The monkey under the stimulus of electricity, makes in a state of unconsciousness a number of grimaces and movements of its arms and legs. These may appear very laughable to some persons, I myself cannot see anything laughable in them."[60]

At first, the Act did not restrain Ferrier's career as he had dreaded. In 1876, he published a popular collection of studies on *The Functions of the Brain*, and in the same year, he was elected a fellow of the Royal Society.[61] In March 1881, Yeo and Ferrier published "The Functional Relations of the Motor Roots of the Brachial and Lumbo-Sacral Plexuses" in the *Proceedings of the Royal Society of London*. The paper criticized the method of anatomical dissection, arguing it was "unable to discriminate between the sensory and motor constituents of the nerve-trunks, or indicate their functional relations and distribution." Physiological research was, therefore, essential, and the only possible method was "excitation or destruction of the individual roots of the plexus." This kind of research was done in "several of the lower animals," but Ferrier and Yeo conducted their research on monkeys,

---

58. David Millett, "Illustrating a Revolution: An Unrecognized Contribution to the 'Golden Era' of Cerebral Localization," *Notes and Records of the Royal Society of London* 52, no. 2 (July 1998): 285.

59. Royal Commission Report, 170; Finger, *Minds behind the Brain*, 165; Finn and Stark, "Medical Science and the Cruelty to Animals Act 1876," 12–23; Leigh Star, *Regions of the Mind*, 5.

60. Royal Commission Report, 169, 83, 171.

61. David Ferrier, *The Functions of the Brain* (New York: G.P. Putnam's Sons, 1876). Ferrier was a student of the Scottish philosopher Alexander Bain, and Gere shows that the book translated Bain's and Herbert Spencer's pain-pleasure psychology into a language of experimental science. Gere, *Pain, Pleasure and the Greater Good*, 153.

"which may be regarded as almost directly applicable to man."[62] It was this kind of research on the sensory and motor function of the brain that led later that year to the earliest and most sensational prosecution under the Act.

One morning in August 1881, physiologists visiting the Seventh International Medical Congress in London met at the Royal Institution on Albemarle Street to discuss the latest developments in brain research. They debated whether the brain could be mapped for physiological functions such as smell, touch, hearing, or movement. Two physiologists were scheduled to present their competing theories on brain physiology.[63] The German physiologist Friedrich Goltz presented first. Goltz questioned the view that the cortex was composed of specialized centers for movements and senses. He described how he "first cut the skull open, and then by the means of a strong stream of water had washed out the grey matter of the brain to a considerable extent."[64] The partial removal of dogs' brains, Goltz argued, led to the diminution of the intellectual powers of the animal, and certain damage to their senses, but generally retained its faculties.[65] The *Nottingham Evening Post* reported how "the interest attaching to the discussion on localization was greatly enhanced by the fact that Professor Goltz had brought one of his dogs from Strasburg."[66] Except for "some clumsiness in movement," Goltz's dog showed "little which would distinguish it from the normal."[67]

Ferrier presented next. He dismissed Goltz's findings and doubted whether Goltz fully destroyed the relevant brain areas. Presenting his own observations on monkeys, Ferrier claimed that the removal of portions of the cortex in the region of the "psycho motor centers" produced paralysis in the limbs of the opposite side of the body. Using a series of microscopic sections of the brain and spinal cord, he argued that the removal of parts of the brain was accompanied by a descending degradation of the nerves supplying the paralyzed muscles.

The demonstration was followed by an invitation to all participants to witness and compare Ferrier's monkeys and Goltz's dog. According to another

---

62. David Ferrier and Gerald F. Yeo, "The Functional Relations of the Motor Roots of the Brachial and Lumbo-Sacral Plexuses," *Proceedings of the Royal Society of London* 32 (1881): 12–20, 14.

63. "The International Medical Congress," *British Medical Journal* 2, no. 1084 (October 8, 1881): 587–98, 588. On earlier stages of the debate over brain localization and its relation to phrenology, see Judith P. Swazey, "Action Propre and Action Commune: The Localization of Cerebral Function," *Journal of the History of Biology* 3, no. 2 (1970): 213–34.

64. "Vivisection," *Bury and Norwich Post, and Suffolk Herald*, November 8, 1881, 3, in 19th Century British Library Newspapers: Part II.

65. E. Makins, G. Henry, and W. MacCormac, *Transactions of the International Medical Congress, seventh session, held in London, August 2d to 9th, 1881* (London: J. W. Kolckmann, 1881).

66. "Charge against a Professor under the Vivisection Act," *Nottingham Evening Post*, November 4, 1881, 4.

67. *"Police Intelligence,"* *Morning Post*, November 4, 1881, 7.

account, the invitation was extended by Yeo—who was in charge of the physiological laboratory at King's College—and arranged by Michael Foster, who was the president of the Physiological Section of the International Medical Congress. Foster later testified that notwithstanding the congress's policy to avoid "being entangled with what is popularly called the vivisection question" and prevent what might seem like a live demonstration of experiments, the overwhelming demand of physiologists to see Ferrier's and Goltz's animals led the congress to relax its rules and show the vivisected animals.[68]

In the afternoon, the scientists gathered at the physiological laboratory of King's College. Ferrier first showed his audience a monkey paralyzed in one of its forelegs. He pinched the monkey in the paralyzed limb and let it walk around to demonstrate its paralysis. But the real thrill was yet to come. Ferrier took out his pistol and approached the second monkey. A shot was heard, but the monkey remained still; the sound did not excite any response. Ferrier explained that he had succeeded in damaging the monkey's hearing by destroying a specific part of its brain.[69] The spectators were astonished by Ferrier's demonstration and eager to investigate its causes. In response to the desire expressed by "a large number of physiologists," both the dog and the paralyzed monkey were killed by chloroform, and their brains were removed to be examined by a committee composed of Edward Schäfer, Emanuel Klein, William Gowers, and John Newport Langley.[70]

While the scientific press celebrated the achievement, the Victoria Street Society was less enthusiastic. Since the Home Office refused to allow the society access to submitted certificates, antivivisectionists meticulously scrutinized scientific publications for evidence of breaches of the Act. The society routinely reported its findings in its journal, the *Zoophilist*, and urged the officials to take action.[71]

---

68. "The Charge against Professor Ferrier under the Vivisection Act: Dismissal of the Summons," *British Medical Journal* 2, no. 1090 (November 19, 1881): 836–42, 840.

69. "The International Medical Congress: Section II—Physiology," *Lancet* 1182, no. 3025 (August 20, 1881): 327.

70. "The Charge against Professor Ferrier under the Vivisection Act," 840.

71. Stephen Coleridge, "The Administration of the Cruelty to Animals Act of 1876," [September?] 1900, HO 144/606/B31612, 394. For example, Stephen Coleridge to Chas. S. Murdoch, February 11, 1899, HO 144/419/B25696-21. Susan Lederer examines the impact similar tactics had on American editing policies. Susan E. Lederer, "Political Animals: The Shaping of Biomedical Research Literature in Twentieth-Century America," *Isis* 83, no. 1 (March 1992): 61–79. For an example of an incident in which Victoria Street Society asked to revoke licenses, see E. Leigh Pemberton to the Secretary of Victoria Street Antivivisection Society, October 5, 1887, HO 156/3, 45. French also comments about the tactic of the society. French, *Anti Vivisection and Medical Science*, 175.

Because the Home Office was reluctant to take public action against physiologists and preferred, since the very first days of the Act, to operate discretely, the Victoria Street Society initiated the first vivisection prosecution. Early in November 1881, the society applied to the Bow Street Court for a summons against Ferrier for violating the Act. The court's hall on the day of the first hearing was populated by supporting scientists, including Yeo.[72] Represented by Q. C. Waddy, Besley, and Bernard Coleridge, the Victoria Society's main argument was that Ferrier did not hold a license to experiment on living animals. In addition, the society argued that Ferrier did not administer anesthesia as required by the Act. Alternatively, the society contended that Ferrier should have killed the monkey as soon as the experiment ended as the law instructed, adding that it was "monstrous to let an animal go about deprived of one or two of its legs."[73] Although the operation on the monkey's brain had long concluded, the prosecutors argued, the animals were suffering from Ferrier's demonstrations. Finally, the society argued that an experiment as an illustration for a lecture was forbidden by the Act, unless under a special certificate. Judge James Ingham granted the summons and scheduled the next meeting for November 17, 1881.[74] The expected penalty in case of conviction was £50, and likely a blow to the reputation of physiological research.

The prosecution alarmed the scientific press, and one reader of the *British Medical Journal* suggested establishing a fund on Ferrier's behalf, to which Charles Darwin also made his contribution to fighting the "absurd and wicked prosecution."[75] Scientific organizations backed Ferrier, and the British Medical Association provided him with legal assistance.[76] On the scheduled day, the parties met again at Bow Street. Unlike the typical antivivisection rhetoric that portrayed vivisectors as heartless torturers, the prosecutors refrained from a personal attack on Ferrier's character and attempted to frame the debate as concerning the principles embodied in the Act. The prosecutors' court speeches included superlatives such as "great men like professor Ferrier," they referred to Ferrier's

---

72. Finger, *Minds behind the Brain*, 171.

73. "Police Intelligence," *Morning Post*, November 4, 1881, 7.

74. "Vivisection."

75. Morton Smale, "The Vivisection Act and Dr. Ferrier," *British Medical Journal* 2, no. 1089 (1881):796; Mr. Darwin to Dr. Lauder Brunton, November 19, 1881, in *The Complete Work of Charles Darwin Online*, ed. John van Wyhe, 2002–, http://darwin-online.org.uk/; Janet Browne, *Charles Darwin: A Biography, Vol. 2—The Power of Place* (Princeton: Princeton University Press, 2003), 486.

76. Ferrier was represented by Q. C. Gully and Mr. Houghton, instructed by Mr. Upton on behalf of the British Medical Association. "The Charge against Professor Ferrier under the Vivisection Act," 836.

"extensive, careful, and no doubt valuable, research," and they emphasized that they did not claim that Ferrier was "a cruel or brutal man."[77]

Waddy, speaking for the Victorian Street Society, argued that the question was not whether Ferrier had performed the operation on the monkey, but rather whether the demonstration was continuous with the operation which was performed six months earlier, "and whether the victims—or as that may be an offensive term, I will say the subjects—of the initial part of the experiment were kept alive by Dr. Ferrier for the purpose of experiments."[78] In that case, the demonstration at the Congress was in contravention of section 3.4 of the Act, according to which "the animal must, if pain is likely to continue after the effect of the anaesthetic has ceased, or if any serious injury has been inflicted on the animal, be killed before it has recovered from the influence of the anaesthetic which has been administrated."

Judge Ingham traced the root of the problem to the definition of experiment. "I do not know how the learned counsel means to define the word 'experiment.' It may be that an actual vivisection has been performed as part of an experiment," he explained, and asked whether "experiment begins and ends with the surgical operation, or does he mean to contend that the experiment is continued over such a reasonable space of time as may attain the object for which the vivisection has been performed?" The definition of an experiment and its duration was also important for a very technical reason. The Act allowed six months to prosecute, starting at the moment of the offense. If the offense was the actual operation upon the monkeys' brains, then the judge had no power over it. Nonetheless, "if the surgical operation was only part of the experiment—if the experiment itself continued for a longer time," then the role of Ferrier in the execution of the operation would have been significant.

Waddy presented the prosecution's understanding of the periodization of the brain localization experiment as follows:

> The experiment was the removal of the brain of the monkeys, one or more, then the careful observation from day to day of the subsequent lives of those monkeys. I believe I am right in saying that there was no benefit to science whatever to be obtained by Dr. Ferrier in this particular instance by the removal of the brains of the monkeys, seeing that the operation was under anaesthetics. The experiment, properly speaking, began after the monkey awoke . . . that is a kind of experiment that is

---

77. "The Charge Against Professor Ferrier," 838.
78. "The Charge against Professor Ferrier," 836.

continued from day to day, and the object is to ascertain whether there is any sensation left.[79]

To support the argument that Ferrier's demonstration was part of an ongoing experiment, Waddy referred to the location of the demonstration in the laboratory and to the social function of consensus to end scientific disputes: "There was a dispute between two eminent medical authorities, which could only be settled by experiment; and they actually adjourned to the laboratory for the purpose of experimenting upon the monkeys and upon the dogs."[80] Waddy also proposed a thought experiment: "Let me suppose that, instead of these animals having the first injury done to them by the hand of Dr. Ferrier, had it done by the hand of any other person, Professor Yeo, for instance." In a hypothetical situation, an animal could be operated upon in a foreign territory not governed by the Act and then brought to Britain for inquiry: "Is it to be safe that any professional gentleman in this country would be entitled to have an operation of that sort performed by an entire stranger on the continent, and then to purchase the injured animal and bring it over to this country to make experiments upon it when it was in a condition to which it could not have been brought in this country in accordance with the law?"[81]

The Victoria Society argued the experiment was "calculated to give pain," although "the question whether there was pain caused to the animals at the time is a matter besides the point." Waddy claimed that section 3.4 of the Act, requiring the killing of an animal before it recovers, referred to pain and serious injury alternatively ("if the pain is likely to continue . . . or if any serious injury has been inflicted,") so even if the monkey was not in pain per se, Ferrier's action was still illegal. Waddy summarized the argument as the following:

> You paralyze the animal so that it cannot feel; it has still the power of motion, but no sensation. You hand it over to an unlicensed person, who says, 'Now it is all ready; it cannot feel pain. But the motor area is still untouched, and therefore the muscles will still work; I may cut off the right hand, and that will not cause pain.' Would that be inflicting

79. "The Charge against Professor Ferrier," 837.
80. "The Charge against Professor Ferrier," 838.
81. Waddy's thought experiment came true a few years later when physiologist Walker asked Erichsen whether he could bring with him back to England animals on which he experimented abroad for class demonstrations, and a few hemophilic rabbits for his own investigation. Lushington replied that it would all depend upon whether the keeping of the animals was an experiment calculated to cause pain, and he advised Walker to ask for both license and certificates. Augustine D. Walker to the Home Office, December 18, 1889, HO 144/316/B7623; Minute, December 21, 1889, HO 144/316/B7623.

a serious injury, although it would not be an experiment calculated to give pain? And can any man contend that that would not be within the meaning of the Act?[82]

But the case took a new turn with the testimony of the physiologist Charles Smart Roy, whose account of Ferrier's demonstration provided the basis for the *British Medical Journal*'s report and—along with the *Lancet*'s report that drew upon notes by Arthur Gamgee—provided the main evidence used by the prosecution.[83] Roy claimed that he had dictated his report to a shorthand writer, only to learn later that the *British Medical Journal* misrepresented his account. Roy testified that as far as he could recollect, Ferrier mentioned brain experiments he had performed on monkeys in the past but did not say that he experimented upon the specific monkeys he presented at the conference. Michael Foster's testimony supported Roy's interpretation that Yeo was the one who operated on the said monkeys.[84]

The question then shifted from the definition of experiment to the distinction between an operator of an experiment, a participant, and a viewer. The prosecution stressed again that Ferrier was responsible under the Act since even if he had not performed the initial brain surgery, his observations constituted an experiment that was "a continuous offence from first to last." But Judge Ingham concluded that the operation was done by Yeo and that Ferrier only inspected the animals, expressing nothing more than "great interest in the result of a cruel experiment performed for the purpose of science no doubt by another person." Following a comment by the defense, the judge changed his phrasing of "cruel experiment" in the above statement to "experiment calculated to give pain." Ingham also encouraged the prosecution to appeal and expressed his wish for "some clear definition of the experiments should be given by a superior court," but it did not happen.[85]

A couple of days after the court session, Yeo delivered a letter to the Home Office, asking to know whether all persons present at an experiment must hold licenses.[86] Yeo phrased his inquiry as a general question and did not disclose his involvement in Ferrier's trial. The Home Office did not explicitly treat it

---

82. Augustine D. Walker to the Home Office, December 18, 1889, HO 144/316/B7623.

83. "The International Medical Congress," *Lancet*, 326; "The Charge against Professor Ferrier," 841.

84. "The Charge against Professor Ferrier," 840. Roy practiced as an assistant to Goltz a few years earlier. "Obituary: Charles Smart Roy, M.A., M.D., F.R.S.," *British Medical Journal* 2, no. 1919 (October 9, 1897): 1031–32.

85. "The Charge Against Professor Ferrier," 841.

86. Yeo to Harcourt, November 19, 1881, HO 144/17/44209F.

as related to the case, although it was doubtless aware of the case. Lushington replied that it was not an offense for a person to be present in an experiment if he did not take part in the performance of the experiment.[87]

Yeo was unsatisfied with the reply and asked the Home Office whether his assistant should hold a license, a question he felt obliged to ask since he was "unable to recognize the difference between a person taking part mechanically and one assisting by his presence."[88] The question was clearly referring to Ferrier's case. Lushington lost his patience. In a note to Home Secretary William Vernon Harcourt, he bitterly commented that Yeo "has before given us trouble about this Act, and would be glad to place the Home Office in difficulties."[89] Lushington might have been right, and Yeo might have intended to embarrass the Home Office with questions about the interpretation of the Act that could put it in opposition to the court. Alternatively, Yeo's experience of Ferrier's execution might have made him cautious, and eager to obtain in advance approval of his planned operations. Lushington suggested answering Yeo with an abstract statement that everyone who conducted an experiment that was calculated to give pain should obtain a license. Nonetheless, the official reply was that the home secretary "must decline to express any further legal opinion on the matter."[90]

Ferrier's trial had a chilling effect on antivivisection societies and became a symbol of the inadequacies of the Act.[91] The trial proved how flimsy the Victoria Society's sources of information were and how dependent the group was on scientists' accounts of their own experiments. The problem of information was demonstrated again ten years later when Home Secretary Herbert Henry Asquith refused a request by the London Anti-Vivisection Society for Home Office papers on Horsley (probably the physiologist Victor Horsley), which needed to prepare a prosecution against him.[92] Ferrier's trial also made clear that the various scientific bodies were quick to back each other up. Frances Power Cobbe blamed the *British Medical Journal* and the *Lancet* for misleading the society with their changing versions of Ferrier's demonstration at the Medical Congress.[93] The

---

87. Memo by Lushington, November 22, 1881, HO 144/17/44209F.

88. Yeo to Harcourt, November 25, 1881, HO 144/17/44209F.

89. Memo by Lushington, November 28, 1881, HO 144/17/44209F.

90. Lushington to Yeo, November 30, 1881, HO 156/1, 404. When asked the same question by physiologist Victor Horsley a few years later, Lushington distinguished between assistants who only handled the instruments and who therefore did not require a license and those actively involved in the operation. Memo by Lushington, November 4, 1889, HO 144/299/B2738A.

91. French, *Antivivisection and Medical Science*, 202.

92. Lushington to the Secretary to the London Anti-Vivisection Society, December 20, 1896, HO 156/7, 388.

93. Frances Power Cobbe, "Vivisection Correspondence," *Zoophilist* 1, no. 8 (1881): 139.

Victoria Street Society did not initiate any other prosecution in the next few decades.

Although physiologists won the case and Ferrier's private practice flourished after the affair, the prosecution remained an unpleasant episode that physiologists wished to avoid.[94] Leigh Star observes that in the trial, "the issues of scientific methods, vivisection medical professionalism, and localization theory became inextricably intertwined in the minds of the profession and the public."[95] The legacy of the case, according to scholar Anne Stiles, was also "the association of cerebral localization with inhumane experimental methods and disturbing philosophical conclusions."[96] The philosophical appeal of Ferrier's research, explains Stiles, was that it demonstrated the similarity between men and beasts, and moreover, that Ferrier's control of the brain using electrical currents defied the sacred place of human will and consciousness.[97]

But in a concrete way, it was the animal rather than the human that lost its will and consciousness. Stimulating the brain to make animals move, scratch, hold, and wave was a new stage in the creation of the animal as a laboratory instrument. In their interpenetration of brain research as an ongoing experiment, antivivisectionists insisted on respecting the integrity of the animal before and after surgical manipulation. But Judge Ingham, lamenting the absence of a clear definition of an experiment, rejected the prosecution's view that the animals presented at the congress were subjects in an ongoing experiment. In the Bow Street Court's view, shared by Yeo in his earlier contention with the Home Office, the mutilated monkeys and dog were no longer animals under experimentation, and thus no longer under the law: once the invasive operation was done, they were modified creatures to be observed for their unique traits.

"Practically a Registration Act": The Establishment of the AAMR

Not long after the Ferrier case was concluded, the Physiological Society delivered a report about the Act, drafted by a committee composed of Lauder Brunton, Philip Henry Pye-Smith, and Gerald Yeo. The committee acknowledged that it

---

94. French, *Antivivisection and the Medical Science*, 202.

95. Leigh Star, *Regions of the Mind*, 56.

96. Anne Stiles, *Popular Fiction and Brain Science in the Late Nineteenth Century* (Cambridge: Cambridge University Press, 2011), 13.

97. Stiles, *Popular Fiction and Brain Science*, 13; Laura Otis, "Howled out of the Country," in Stiles, *Neurology and Literature*.

was "too late to enquire whether or no it was wise policy to allow the Bill to pass in 1876" and that it was "practically useless" to attempt to repeal it. The last resort in opposing the Act would be to return the licenses "and take our chance of prosecution for some well selected 'test' experiments." This sort of scientists' revolt never happened. Instead, physiologists adopted the next recommendation: to turn the Act into "a protection instead of hindrance if reasonably administered."[98]

The Physiological Society's committee advised urging the Home Office to change the administration of the Act, "so as to make it practically a registration act, with the double safe-guard against possible abuse of the 'recommendations,' and of the 'inspection' which it provides." On a practical note, the committee recommended that physiologists "should only apply for such certificates as are necessary for their special purpose, that they should always be prepared, in the event of their case being submitted to referees by the home secretary, to show their scrupulous attention to the provisions of the Act, and that they should not be deterred from sending in any necessary certificate by the chance or the certainty of its being disallowed."[99]

The report contained findings from a survey based on thirty respondents (out of seventy questionnaires sent) showing that there has been "no absolute refusal of a license, but, that refusals of both licenses and certificates have repeatedly been revoked, only after a long delay and by the help of strong pressure upon the Home Office: while some certificates have been absolutely refused. Moreover, in several cases, experiments have been prevented owing to intimation that licenses or certificates would certainly not be granted." The committee reported about seven cases of refusal to allow certificates, six cases caused "injurious delay, often amounting to practical refusal," and five cases in which experimenters were deterred from applying.[100] This account reveals something important about the working of the Act: even though there were no "absolute refusals" of license applications, the Home Office administration of the Act—characterized by polite requests for application revisions and only seldom a refusal—had impacted the way physiologists approached their research. What the Physiological Society described in its report are scientists who feel discouraged from some actions by

98. Report of a Committee appointed by the Physiological Society in pursuance of the following Resolution passed on October 15, 1881 (as amended and adopted by the society at its meeting of the 8th Dec.), n.d., WA, SA/RDS/A/3.

99. Report of a Committee appointed by the Physiological Society in pursuance of the following Resolution passed on October 15, 1881.

100. Report of a Committee appointed by the Physiological Society in pursuance of the following Resolution passed on October 15, 1881.

the threat of administrative action, and their realization that they must find ways to cope with the new legal-moral order at the same time as trying to adjust it to their needs.

The report also advised working on "enlightening the public in general, and Parliament in particular on the nature, objects, and means of physiological and pathological research."[101] On March 28, 1882, a representative meeting of members of the medical profession was held in the library of the Royal College of Physicians. William Jenner, president of the Royal College of Physicians, and Erasmus Wilson, president of the Royal College of Surgeons, had signed the invitation.[102] At the meeting, scientists from various institutions agreed to establish the Association for the Advancement of Medicine by Research (AAMR), which, obscured from public awareness, transformed the administration of the Act.[103]

The first meeting of the council of the AAMR was held at the Royal College of Physicians on April 20, 1882.[104] Following the meeting, the surgeon James Paget delivered a letter to Home Secretary Harcourt, explaining that "one of the objects of the Association is to secure that while the spirit of the Act . . . shall be strictly adhered to, those researches upon which the progress of medical knowledge especially depends may be . . . promoted." The association proposed its "aid or advice" in administrating the Act, "whether you would be willing to avail yourself of this special knowledge in your administration of the law." The AAMR also offered to submit comments on the workings of the Act and "suggestions to changes without contradicting the law."[105]

In June 1882, Lushington informed the AAMR that Home Secretary Harcourt would be "glad to avail himself to their advice and assistance."[106] Harcourt's move unveiled a bias in favor of the scientific lobby, but it also reflected his wish to

---

101. Report of a Committee appointed by the Physiological Society in pursuance of the following Resolution passed on October 15, 1881. The structure of the future organization was detailed in Draft for Purposed Resolution, n.d., WA, SA/RDS/A/3.

102. A letter from William Jenner and Erasmus Wilson, March 14, 1882, WA, SA/RDS/A/3.

103. French, *Antivivisection and Medical Science*, 204–5; Turner links the establishment of the AAMR to the Ferrier trial. Turner, *Reckoning with the Beast*, 108.

104. Minute Book of the Association for the Advancement of Medicine by Research, April 20, 1882, WA, MS/5310, 3. At the first meeting of the executive committee of the AAMR council, hosted by James Paget in his home, they discussed a suggestion that the association will found a laboratory. Minute Book of the Association for the Advancement of Medicine by Research, May 2, 1882, WA, MS/5310, 5.

105. Copy of a letter from James Paget to Harcourt, Minute Book of the Association for the Advancement of Medicine by Research, May 1882, WA, MS 5310, 5.

106. Copy of a letter from Lushington to the AAMR, June 6, 1882, Minute Book of the Association for the Advancement of Medicine by Research, WA, MS 5310, 14.

ease the ever-increasing workload at the Home Office. In 1880, the annual number of incoming papers was estimated at 44,541, an increase of 48 percent from 1872. As Pellew observes, "the department reacted to pressure by trying to have responsibilities removed from it wherever this was politically acceptable."[107] In December 1882, Harcourt offered the AAMR the following arrangement: the Home Office would forward each application to a member of the association who would present it to the AAMR council. Once recommended by the council, the Home Office would consider the application.[108] Unsurprisingly, the arrangement with the AAMR, which remained in force until 1913, drew criticism from antivivisectionists. Years later, the International Anti-Vivisection Council referred to the arrangement as "injustice which must be repugnant to every fair minded person."[109]

The AAMR appointed a subcommittee to be responsible for processing the applications for licenses and certificates, stipulating that each decision would be presented to the executive committee for approval before being sent to the Home Office.[110] The number of applications refused by the AAMR between the years 1882 and 1890 was negligible.[111] However, things did not go as smoothly as the AAMR had hoped, and as the flowing sections would show, the Act did not easily turn into a mere "registration act."

The main issue of contention after 1880 was not whether to regulate vivisection, but how. The medical lobby adopted tactical conformity with the Act's requirements while strategically attempting to increase the influence over the administration of the law. In parallel, the medical lobby invested in what Rob Boddice calls the development of "public relations machinery," propagating the legitimacy of the experimental project.[112] This included not only zealous publications of the therapeutic potential of animal research but also invalidating prevalent assumptions about animal pain and questioning its accurate interpretation by the Home Office and the public at large.

---

107. Pellew, *The Home Office*, 39.

108. Copy of a letter from Lushington to Philip Pye-Smith, December 30, 1882, Minute Book of the Association for the Advancement of Medicine by Research, WA, MS/5310, 22.

109. Honorary Secretary of the International Anti-Vivisection Council to Herbert Gladstone, April 20, 1906, HO 144/964/B18250.

110. Minute Book of the Association for the Advancement of Medicine by Research, WA, MS/5310, 8; French, *Antivivisection and Medical Science*, 207.

111. French showed that the in the only three cases of "outright and final refusals" the applicants had no institutional affiliation. French, *Antivivisection and Medical Science*, 208.

112. Rob Boddice, *Humane Professions: The Defence of Experimental Medicine, 1876–1914* (Cambridge: Cambridge University Press, 2021).

Fact v. Fancy

In April 1883, a short book titled *Physiological Cruelty: Or, Fact v. Fancy—An Inquiry into the Vivisection Question* was published anonymously. The author, Philanthropos, aimed to provide an "unprejudiced investigation" into the vivisection controversy, although he clearly approved of vivisection and was critical of the Act.[113] George J. Romanes, a founding member of the Physiological Society, wrote an enthusiastic review of this piece for *Nature*, which praised Philanthropos's medical knowledge and speculated that he was a working physiologist.[114] Romanes was right, and the writer was none other than Yeo, as was revealed by his proposal to contribute three hundred copies of "his book 'Physiological Cruelty'" in exchange for a reduction in his AAMR membership fees.[115]

Philanthropos began his inquiry with the question "What is pain?" replying that pain is a personal experience, and "as a matter of fact, we know nothing about any pain except what we have ourselves suffered." Humans have learned to grasp each other's pain, but "we lose ourselves at once" when confronted with the pain of animals. He claimed that "signs of pain" produced by animals were vague and neither proved the existence of consciousness nor disclosed the degree of feelings. Moreover, because of the reflex action, "motions, cries, jerks, and struggles" should not be considered reliable signs of pain. To conclude, "it is not safe to judge of what is terrible to suffer by what is terrible to witness."[116]

Lay attempts to empathize with animals were useless from this perspective. A year earlier, Richard Holt Hutton, a journalist and theologian who served on the Royal Commission on Vivisection of 1875, published a plea for empathy in the *Nineteenth Century*, stating, "I think that in a rough way we may put ourselves in the place of the lower animals, and ask what we, with their pains, and their sensitivities, and their prospects of life, and pain, and happiness, might fairly expect of beings of much greater power, but of common susceptibilities."[117] In response, Philanthropos contended "put yourself in his place argument"

---

113. Philanthropos. *Physiological Cruelty, Or, Fact v. Fancy: An Inquiry into the Vivisection Question* (New York: John Wiley and Sons, 1883), B.

114. George J. Romanes, "Physiological Cruelty, or Fact versus Fancy; an Inquiry into the Vivisection Question," *Nature* 28, no. 727 (October 4, 1883): 537. More on Romanes and his view of vivisection in Rob Boddice, "Vivisecting Major: A Victorian Gentleman Scientist Defends Animal Experimentation, 1876–1885," *Isis* 102, no. 2 (June 2011): 215–37.

115. Minute of Executive Committee, December 23, 1885, WA, MS 5310, 45. See also French, *Antivivisection and Medical Science*, 198.

116. Philanthropos, *Physiological Cruelty*, 4, 5, 7, 17.

117. Richard Holt Hutton, "The Biologists on Vivisection," *Nineteenth Century* 11 (1882): 38.

demonstrated "the confusion between what animal cannot possibly understand, and what we should have to suppose ourselves in their place understanding, in order to pass judgment of their treatment—is inextricable, and altogether it is an idea that cannot be worked out."[118] Pain, therefore, could not be detected by the untrained eye. It was a task for the knowledgeable—those who are immersed in the medical professions—to discern a genuine pain. In this anonymous publication, Yeo discredited the antivivisectionists' as well as the Home Office's empathic skills and authority to identify animal pain.[119]

Under a pseudonym and without disclosing his ties to the case, Yeo used the Ferrier trial as an example of lay miscomprehension of animal pain. Philanthropos complained about the bad reputation of brain surgeries, while "as a matter of fact, injury to the brain itself causes *no* pain." Yeo provided an example from the German medical periodicals of a young man who was injured, and "a considerable quantity of the brain-substance escaped on three several days." The man's wound was sterilized, and despite "many paralytic symptoms," which ultimately healed, "*he never complained of any pain, nor even had the least headache.*" He concluded that "injuries to the brain are painless to men, and must, therefore, be painless to animals," given that animals suffer less than humans, due to their inferior consciousness and intellect. The only pain was "that to the feelings of tender-hearted people ignorant of physiology."[120]

The *Zoophilist* responded to the publication of *Physiological Cruelty* by quoting a description of Ferrier's monkeys published in the *Lancet*, wondering: "And 'Philanthropos' sees no signs of suffering in these maimed and mutilated creatures; nothing to pity; nothing to execrate?" The competing views of Yeo and the *Zoophilist* sharpened the differences in the attitudes and understandings that had been expressed in Ferrier's trial. One perspective focused on a narrow conception of pain, pointed to the medical technologies that could nullify it, and rejected empathetic thought experiments while relying on the testimony of humans undergoing similar procedures. The other perspective incorporated mutilation and other damages to the animals' capacities into its definition of pain, avoided physiological terminology, and scorned those who were "dead to all sympathy."[121]

---

118. Philanthropos, *Physiological Cruelty*, 38.

119. About the tensions between physiologists' dismissal of animals' expression of pain and Darwinian evolutionary theory, see Jed Mayer, "The Expression of the Emotions in Man and Laboratory Animals," *Victorian Studies* 50, no. 3 (2008): 399–417, 403–4.

120. Philanthropos, *Physiological Cruelty*, 18, 19.

121. "Notices of Books: Physiological Cruelty," *Zoophilist* 2, no. 10 (1883): 161.

Back in the realm of bureaucracy, Yeo still needed his forms to be signed. November 1883 approached, and Yeo had to submit his annual request for license renewal. He included an application for a certificate B (allowing the experimenter to recover the animal from anesthesia rather than killing it) signed by John Marshall, the president of the Royal College of Surgeons. The aim of Yeo's planned experiment was "to investigate the result of a lesion of certain parts of the cerebral cortex of dogs. Always under chloroform during the operation" and treated in a way aiming "to minimize pain."[122] The AAMR sent a supporting letter to the Home Office, stating that after learning from Yeo "the mode in which the proposed experiments would be conducted," it advised that the certificate would be allowed for three experiments.[123] Inspector George Busk had "no hesitations" in concurring with the recommendations of the AAMR. He considered the experiments "not necessarily painful," of "great scientific value," and likely to contribute to the advancement of the knowledge of the cerebral function. Yet "in lieu of the public feeling" evidenced in "the attack . . . upon Dr. Ferrier," Busk recommended restricting these kinds of experiments as much as possible.[124]

The potential public critic was also a matter of concern to Lushington, who was reluctant to approve Yeo's request. In a detailed memo to Harcourt, he informed the home secretary about the AAMR's endorsement of the application and explained inspector Busk's favorable view on the topic. "You will probably be inclined to allow their recommendation," predicted Lushington, "but you should be made aware that these experiments when they become known, are certain to provoke a good deal of public comment." First, because they were severe experiments that will keep the animal alive "for a considerable time in considerable pain or at least distress." The second reason was simply because "the animals will be dogs."[125]

A third reason for denying Yeo's certificate application was the changing legal landscape. Lushington admitted that in previous years, the Home Office had not asked Yeo to submit, in addition to the other forms, certificate E as required for experimenting on dogs and cats without anesthesia. However, recent opinions of the Law Officers regarding the inoculation experiment (the Greenfield Case of 1879 and the Roy Case of 1882 discussed in chapter 3) required the home secretary to readdress the question of whether keeping an animal alive could be termed an experiment calculated to give pain. The Law Officers ruled in these

---

122. Yeo, Form of Certificate B, November 8, 1883, HO 144/17/44209F.
123. AAMR [Clinton Dent?] to Lushington, November 14, 1883, HO 144/17/44209F.
124. Busk to the Secretary of State, November 28, 1883, HO 144/17/44209F.
125. Memo by Lushington, November 28, 1883, HO 144/17/44209F.

two cases that an operation that involved keeping animals alive after inoculation was an experiment calculated to give pain and therefore governed by the Act. The same logic, reasoned Lushington, must be employed while considering Yeo's application.

Remarking on the resemblance of Yeo's planned experiments to those for which Ferrier was prosecuted, Lushington stressed that it was "very necessary at all events to be perfectly sure that we are on firm legal ground and this I very much doubt." It was, therefore, necessary this time for Yeo to submit certificate E, "clarifying that the effects of the experiment would be frustrated if performed on any other animal than a dog." Lushington explained that the "terms of the Statute are very ambiguous" and advised the home secretary to ask for clarifications from the Law Officers.[126] However, the case, it seems, was not forwarded to the Law Officers.[127] Instead, Home Secretary Harcourt asked the AAMR for a statement explaining why they recommended that the experiment be allowed and also "why in their opinion these experiments must be performed on a dog."[128]

Yeo did not pursue his objection to submitting certificate E any further, and in late December 1883, he delivered it to the Home Office. In the certificate form, Yeo described the controversial experiment as an inquiry into the "effects of lesion of the motor center of the cortex of the brain of the dog." Under the rubric of "the reasons why the experiment will be frustrated if operated on a different species," Yeo wrote: "Because the brain of the animals are not exactly the same as that of the dog, and the comparison with the results of other animals can only be made by operating on these animals."[129]

On the margins of the certificate form, Yeo added a note: "The operation will be conducted while the animal is under the influence of chloroform but as some think the mere observation of the animal after the effects of the anaesthesia has passed off is an experiment under the statute I apply for this certificate, which I shall apply only for three experiment observations" (figure 3).[130] In the attached support letter by the AAMR, the organization cited Yeo's note and emphasized that although Yeo submitted certificate E, he was not proposing to perform the experiments without anesthetics.[131] They did not, however, explain why dogs were indispensable for Yeo's proposed experiment. Harcourt

126. Lushington to Harcourt, November 28, 1883, HO 144/17/44209F.

127. Extract from Minute on Dr. Yeo Papers, November 1883, HO 144/16/44209.

128. Memo by Lushington, December 1, 1883, HO 144/17/44209F; Lushington to the Secretary of the Medical Research Association, December 4, 1883, HO 156/1, 672.

129. Gerald Yeo, Certificate E, n.d., HO 144/17/44209F.

130. Gerald Yeo, Certificate E, n.d., HO 144/17/44209F.

131. Clinton Dent to Lushington, December 20, 1883, HO 144/17/44209F.

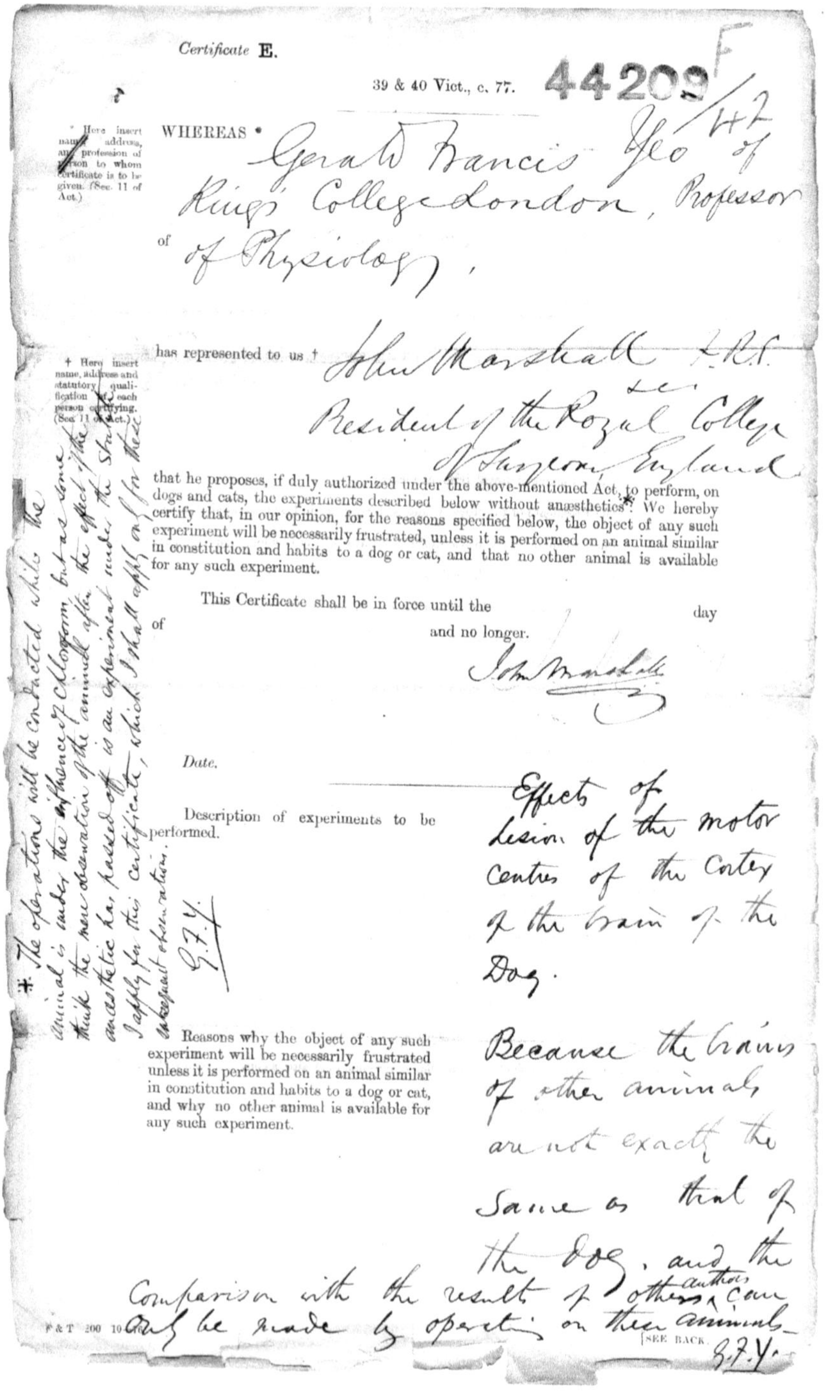

FIGURE 3 / Gerald Yeo's certificate, with a note on the margins. Certificate E, n.d., TNA, HO 144/17/44209F.

instructed Lushington to withhold the decision until after they received a letter from the physician William Jenner, president of the Royal College of Physicians, on an apparently related topic.[132]

The Home Office made no decision for two weeks after Yeo submitted his application. In early January 1884, the AAMR sent a letter to Lushington, urging him to approve Yeo's certificate. The association claimed that Yeo's experiments aspire to "elucidate a point in physiology" in which "some light has been thrown already by physiological experiment" but was still obscure. It also contended that there were "diametrically opposite opinions" on the subject of research, which Yeo's experiment might settle. The AAMR acknowledged that the exploration of the cortex was not of a "pressing interest" but rather "a fundamental one" in the field of physiology, adding that "the experiments are proposed because the force of reason can go no further in the matter."[133] The Home Office postponed its reply again.[134]

Members of the AAMR were enraged by the way the Home Office handled Yeo's request. Yeo was among the founders of the association and its provisional secretary in its very first days, a member of its executive committee, as well as a member of the subcommittee that prepared a report on the administration of the Act.[135] Debating its strategy about Yeo's unapproved application, the executive committee of the AAMR decided to maintain the support of Yeo yet avoid taking any further action.[136] In a meeting of the Physiological Society at King's College, about a week later, Yeo complained that his work had been obstructed by the Home Office.[137] During the same month of January 1884, Yeo completed his *Manual of Physiology: A Text Book for Students of Medicine*, which received positive reviews and was published in several editions, including in the United States.[138]

Despite the AAMR's firm and united stand when communicating with the Home Office, not all members of the organization held the same view about

132. Memo, December 20, 1883, HO 144/17/44209F.

133. Clinton Dent to Lushington, January 5, 1884, HO 144/17/44209F; Memo by Lushington, January 5, 1884, HO 144/17/44209F.

134. Memo by Lushington, January 7, 1884, HO 144/17/44209F.

135. Minute Book of the Association for the Advancement of Medicine by Research, April 20, 1882, P.3, WA, MS/5310, 3; Minute Book of the Association for the Advancement of Medicine by Research, May 2, 1882, 5.

136. Minute of the Executive Committee of the AAMR, January 11, 1884, WA, MS/5310, 32.

137. Sharpey-Schafer, "History of the Physiological Society during Its First Fifty Years," 75.

138. "Review of a Manual of Physiology, for the Use of Junior Students of Medicine, by Gerald F. Yeo," *British Medical Journal* 1, no. 1217 (1884): 820–21; "Review of a Manual of Physiology. A Text-Book for Students of Medicine, by Gerald F. Yeo," *Science* 11, no. 273 (1888): 204.

experimentation on dogs. In February 1884, the executive committee discussed the subject of certificate E and experiments on domestic animals. William Jenner stated at length the grounds on which he had considered it inadvisable to sign a certain certificate E that he was asked to approve, explaining that he was unwilling to support any application for using dogs for experimentation unless he was convinced of the "absolute necessity" of using these animals. In the specific case he was addressing but its details were not disclosed, it appeared to him the experimenter's chief arguments in favor of using dogs given by the applicant were those of "convenience and economy." At that moment, the secretary of the meeting reminded his colleagues that Yeo's application for a similar certificate E was still under consideration by the Home Office. The physician Philip Pye-Smith spoke strongly in favor of pressing the home secretary to grant the application. After a considerable discussion, the executive committee resolved to transfer the topic of experimenting on domestic animals to a special subcommittee.[139]

Weeks passed, and Yeo received no answer from the Home Office. The AAMR delivered another letter, asking for the renewal of Yeo's license, "in order that he may not be hindered in the ordinary physiological work," writing nothing about the controversial certificates.[140] Finally, in March 1884, four months after Yeo submitted his requests for license renewal and certificates B and E, the Home Office renewed his license. However, it informed Yeo that "under the circumstances," it decided to suspend the certificates."[141] Yeo sent two successive letters, again requesting his certificates, and asking for an explanation of the rejection.[142] In an effort to alleviate anger, Lushington explained that the certificates were suspended pending inquiry rather than disallowed.[143]

In the same month of March 1884, the AAMR subcommittee on domestic animals completed its mission. The association sent the home secretary a memorial approved by its council, of which Yeo was a member. The AAMR complained that certificate E was not explicitly recommended by Royal Report and that the kind of experiments which required certificate B (letting the animal recover from anesthesia) "with very rare exceptions, entail no more suffering than that of a patient during convalescence after a surgical operation conducted with modern precautions against inflammation."[144] In other words, the AAMR contended that

139. Minute of the Executive Committee of the AAMR, February 15, 1884, WA, MS 5310, 33.
140. Dent to Lushington, March 10, 1884, HO 144/17/44209F.
141. Lushington to Yeo, March 13, 1884, HO 156/2, 26; Lushington to AAMR, March 13, 1884, HO 156/2, 27.
142. Yeo to Lushington, March 14, 1884, HO 144/17/44209F; Yeo to Lushington, March 27, 1884, HO 144/17/44209F.
143. Memo by Lushington, March 29, 1884, HO 144/17/44209F.
144. Memorial to the Rt. Hon. The Secretary of State for the Home Department, Unanimously Approved at a Meeting Held March 18, 1884, WA, SA/RDS/A/3, 1, 4.

the only painful aspect of procedures such as brain surgery was the recovery from anesthesia.

The AAMR's council expressed its willingness "to adopt any further precautions which may seem desirable to the Home Office, in order to ensure that dogs shall not be used in any cases in which other animals can be substituted," as the association's members "appreciate and to a large extent share the special sentiments of the public on this point." Additionally, the AAMR was engaged in looking for animals that could serve as substitutes, "to which these sentiments would not in the same degree apply."[145]

The use of dogs, nevertheless, could not be completely avoided. The AAMR then proposed considering the use of dogs condemned to be destroyed under the rabies legislation, the Act to Provide Further Protection against Dogs (1871). Inevitably, the question of Yeo's application came up. The council reasserted its support of Yeo's research, emphasizing its importance for the relief of human suffering and the advancement of knowledge and claiming that it "cannot be carried on without the use of dogs" and that "the pain or discomfort experienced by these animals will be very small indeed." The council concluded the document with direct criticism of the home secretary for its lack of support in medical research.[146]

The council's memorial was supplemented by a "Memorandum of the Reasons which determine the Use of Domestic and other Animals for Purpose of Physiological and Medical Investigations." The memorandum included a survey of the animals subjected to experiments: frogs and other cold-blooded animals; pigeons; and warm-blooded animals such as rodents, cattle, domestic carnivores, horses, and monkeys—the latter were "of the highest values in respect of observation upon the brain and of certain questions in pathology; their intelligence and highly developed senses made physiologists unwilling to use them for any purpose involving pain, and public sentiment would probably fully participate in this reluctance."[147] As for cats and dogs, the memorandum cited the findings of the 1876 memorandum composed by British physiologists, claiming they were indispensable for "a considerable number of researches."[148]

The AAMR continued to pressure the Home Office regarding Yeo's application. In the letter attached to the memorial and the memorandum, AAMR secretary Dent asked to send a deputation consisting of the president of the AAMR

---

145. Memorial to the Rt. Hon. The Secretary of State for the Home Department, 5.
146. Memorial to the Rt. Hon. The Secretary of State for the Home Department, 6.
147. Memorandum of the Reasons which determine the Use of Domestic and other Animals for Purpose of Physiological and Medical Investigations WA, SA/RDS/A/3, 8.
148. Memorandum of the Reasons which determine the Use of Domestic and other Animals for Purpose of Physiological and Medical Investigations, 9.

Spence Wells and other leading physiologists such as Thomas Henry Huxley and Joseph Lister, to discuss his decision in Yeo's application and to "uphold generally the opinions and feeling expressed in the 'memorial and memorandum.'"[149] The home secretary declined the request for a meeting.[150]

Deprived of the option to experiment on dogs and cats, Yeo continued his collaboration with Ferrier in experimenting on monkeys. In April 1884, they published "A Record of Experiments on the Effects of Lesion of Different Regions of the Cerebral Hemispheres" in the *Philosophical Transactions of the Royal Society of London*. The publication integrated some of their earlier findings, including Ferrier's demonstration at the International Medical Congress in 1881. Ferrier and Yeo explained in their introduction that the monkeys were "thoroughly narcotised with chloroform, and kept in a state of complete anesthesia during the whole of the operative procedure."[151] They made the lesions in various areas of the brain using mainly the method of galvanic cautery, a practice thought to help prevent infections.

In the following years, physiologists increasingly used monkeys while the Home Office unofficially extended the Act's special provisions to include them. Henry Matthews was appointed home secretary in August 1886. In November, he asked that the inspectors specify in their reports the probable amount of pain resulting from experiments involving certificates A, B, E, and F (experiments without anesthesia; experiments allowing the animal to recover; experiments without anesthesia on dogs and cats; and experiments on horses, mules, and asses). Matthews also instructed that any case in which a considerable amount of pain might be inflicted, or which involved cats, dogs, or monkeys, should be brought to his personal notice.[152] In 1893, Asquith, Matthews's successor at the Home Office, exhibited the same commitment to oversee experiments on the "higher animals":

> All I have got to do is to administer the Act as I find it; and I may say that there is no part of my duty that gives me greater solicitude, or to which I pay, as time allows, more constant attention. In no case, I think, since I have been in the Home Office has a license to experiment upon dogs,

---

149. Dent to Harcourt, April 2, 1884, HO 144/17/44209F; Memo, April 2, 1884, HO 144/17/44209F.

150. Note, n.d., HO 144/17/44209F; Liddell to the Secretary of the Medical Research Association, April 1884, HO 156/2, 32.

151. David Ferrier and Gerald F. Yeo, "A Record of Experiments on the Effects of Lesion of Different Regions of the Cerebral Hemispheres," *Philosophical Transactions of the Royal Society of London* 175 (1884): 479–564, 480.

152. Lushington to Stapleton, November 24, 1886, HO 144/17/44209X.

cats, monkeys, and the higher class of animals been granted without the matter having come before me personally, and receiving my individual attention.[153]

From Science to Law and Back: The Antiseptic Condition

In January 1888, Home Office inspector John Eric Erichsen composed a "Memorandum Respecting the Use of Anaesthetics and the Employment of Antiseptics in Experiments on Living Animals."[154] The memorandum was concerned with the animals protected under certificates E and F (dogs, cats, horses, asses, and mules) with the addition of monkeys. It opened with the Act's requirement to use anesthetic when pain is likely to occur during an experiment. Since the Act had passed, "a considerable advance has taken place in the use of anaesthetics and in the treatment of wounds. Certain local anaesthetics are now used, such as cocaine and carbolic acid, which completely deaden the sensibility of the most sensitive structures" such as the eye and the skin. A new promising practice was the "antiseptic method," with which, according to Erichsen, "the pain of a wound is materially diminished."[155] Yeo was one of the contributors to the study of antiseptics.[156]

Erichsen explained that it was "universally recognized by surgeons" that there were "two elements in the production of pain in a wound and two periods, at which it occurs." One kind of pain was produced by the instrument that inflicted the wound, and the other was caused by subsequent inflammation and suppuration of the wound. The first kind of pain can be "absolutely prevented by the use of an anesthetic of sufficient power to produce unconsciousness or by certain local anaesthetics." However, when the animal recovered from the anesthesia or the effect of the local anesthetic had worn off, the second type of pain appeared. If a wound was "treated by the old and unscientific method," pain was inevitably produced by "the exposure to the air and the development of inflammation in the wounded part." By the "modern method, dressings are applied before the

153. 17 Parl. Deb. (4th ser.) (1893) 343.

154. John Erichsen, Memorandum Respecting the Use of Anaesthetics and the Employment of Antiseptics in Experiments on Living Animals, January 21, 1888, HO 144/299/B2719.

155. Erichsen, Memorandum Respecting the Use of Anaesthetics, 2.

156. Gerald F. Yeo, "Note on the Application of the Antiseptic Method of Dressing to Craniocerebral Surgery."

animal recovers consciousness, as the air is excluded from the wound no inflammation or suppuration ensues and no pain is experienced."[157]

Erichsen claimed that the anesthetic qualities of antiseptics were proved by "the daily experience of ordinary surgical practice" on humans and argued that the substance had the same effect on the "lower animals," since after severe operations such as the removal of tumors in canines and in veterinary practice, "no sign of suffering is manifested by the animal" when treated with antiseptics. With the use of antiseptics in surgical experiments, "the animal may be allowed to recover consciousness, and to live with the certainty that no pain will be suffered, so long as the antiseptic dressing hold good. Should they, by any accident be displaced or fail, and inflammation be set up in the wound with its attendant pain, then the animal should be destroyed."[158] Erichsen concluded:

> In many cases therefore it would be more merciful to dress the wound antiseptically and to allow the animal to live, than to destroy it before it had recovered consciously, with the view of saving it from pain which it would no doubt have formally suffered, but which does not occur under the improved methods of treatment already known as the "antiseptic," provided always that the experiment had not entailed such a mutilation as to interfere permanently with the comfort and enjoyment of the animal.[159]

Erichsen's report adopts the claim, made by Yeo in his certificate applications, that an animal treated rightly for its wounds could recover from the operation painlessly. He is enthusiastic about the pain relief offered by the new medical tool, and in this sense, he sides with physiologists' claim that they are able to produce the means to control pain in laboratories without legal oversight. However, Erichsen added a reservation regarding mutilated animals—such as those that Yeo and Ferrier experimented on—whose "comfort and enjoyment" were compromised despite the use of disinfectants. Following the memorandum, a condition requiring experimenters to use antiseptic was inserted into licenses to ease the pain of animals recovering from experiments under certificate B.[160] The antiseptic condition dictated a certain medical treatment for animals subjected to surgery, thus deepening the Home Office intervention with physiological practice.

---

157. Yeo, "Note on the Application of the Antiseptic Method of Dressing," 3.
158. Yeo, "Note on the Application of the Antiseptic Method of Dressing," 3.
159. Yeo, "Note on the Application of the Antiseptic Method of Dressing," 5.
160. Second Royal Commission, *Final Report*, 7.

The flow of information about animals' bodies went back and forth between the Home Office and physiologists. That flow also carried moral values. As demonstrated in Yeo's essays, physiologists incorporated the discourse on animal well-being. Yeo opened his 1881 paper on antiseptics with a note about how he treated his monkeys, the same note he later used in his application for a special certificate. A few years later, Erichsen prepared the above report based on scientific data about antiseptics and their contribution to the demolition of pain. The circular movement of knowledge about animal pain and rhetoric of animal care was completed when the new scientific knowledge about antiseptic and pain led the Home Office to make the use of antiseptic mandatory in scientific practice.[161]

The availability of anesthesia allowed vivisection practitioners to deny that pain produced a special problem in the laboratory. An 1882 AAMR memorandum provided that "Happily, the amount of pain inflicted in the course of scientific experiments need only be small, and the destruction of life insignificant," because "science herself provided the means by which pain is reduced to a minimum." It was primarily thanks to the "beneficent discovery of anesthesia" which was "one cause of the great difference between the suffering inflicted by Harvey, Boyle, Hales, Haller, Hunter, Magendie, and Bell, and the generally painless experiments of a modern laboratory."[162] But the benefits of anesthesia, shared by both pro- and antivivisection people, was what singled out brain research as problematic.

During the Ferrier trial discussed above, the prosecutor reasoned that according to his idea of the Vivisection Act, "there is no experiment provided for by this statute as being lawful which does not begin with anaesthetics and end with death."[163] This statement captured the suppositions undergirding the Act and its shortcomings, as an increasing number of experiments neither started with anesthesia nor necessarily ended with death. The Home Office responded to the challenges posed by brain research with the resolution to issue special certificate E for these kinds of surgical experiments.[164] This policy led to a surge in the number of experiments in the annual reports designated as done without anesthesia. At the same time, however, these experiments were classified in the reports as painless, leading antivivisectionists to protest that "experiments were put down as painless which had been carried on under antiseptics. Undoubtedly

161. Copy of License no. 693, December 31, 1899, HO 144/445/B30018.

162. AAMR, Memorandum of Facts and Considerations Relating to the Practice of Scientific Experiments on Living Animals, Commonly Called Vivisection, June 1882, WA, SA/RDS/A4, 5.

163. "The Charge against Professor Ferrier under the Vivisection Act," 841.

164. Minutes, [26?] October 1898, HO 144/419/B25696-21.

at the beginning the experiments were painless, but then came the awakening with all its prolonged agonies," and therefore, "the cruelties of vivisection were not prevented under the Act of 1876."[165]

The debates surrounding the legal classification of brain research demonstrate how the implementation of the Act was a process of simultaneously producing pain and experiments. Since it governed "any experiment calculated to give pain," deciding when experiments ended meant setting the limits of the law and thereby excluding some kinds of pain from its consideration. Attempts by the Home Office, the justice system, and physiologists to define an experiment necessitated them to tackle the definition of pain, and vice versa: as animal pain was a matter of concern only if it was inflicted during an experiment, attending to pain required first determining the context in which pain was produced. A similar dilemma was raised in the context of inoculation experiments, which involved more experiments, many more animals, and a few law officers. The disturbing question mark that hovered above the concept of experiment in the brain research debate traveled to that of pain in the inoculation controversy while keeping the two questions interlocked.

---

165. 52 Parl. Deb. (4th ser.) (1897) 373.

3

THE PRICK OF A NEEDLE /
The Challenges of Inoculation

The scientific understanding of infectious diseases changed dramatically during the nineteenth century. Miasma theories of disease, which attributed the generation of disease to environmental factors such as dirt and smells, gave way to a search for specified disease agents. Epidemiology was a medico-legal field from its outset when each epidemiological theory was matched with a different, albeit sometimes overlapping, stately response. When miasma theories had the lead, the British harnessed public sanitary initiatives to eradicate diseases. When the medical profession's consensus tilted at mid-century toward germ theories of disease transmission, the central government launched public vaccination campaigns (at first voluntarily, then compulsory) and, in the last decades of the century, oversaw epidemiological research under the Vivisection Act.[1]

The surgeon Edward Jenner introduced the smallpox vaccine at the dawn of the eighteenth century. Inoculation, the induction of resistance to an infectious disease by introducing a disease-contaminated matter into a healthy body (also referred to as "variolation"), was well known in Jenner's time. Jenner's novelty lay in his harnessing of cowpox, which manifested mildly on human bodies, to defeat the deadly smallpox. This vaccination method interlocked human and animal medical knowledge, practices, and bodies. The interspecies mixture invoked resistance from some, feeding on the revulsion from the insertion of beastly materials into humans. However, Jenner's later innovative techniques of human-to-human transfer of cowpox and the use of dried infectious matter to speed up the vaccine's global distribution, reduced the need of diseased cows in the process (although, as Boddice notes, cows' presence was always implicit).[2]

Real animals reappeared forcefully in the bacteriological research of the 1880s, fulfilling various roles; they were used as vessels of infectious agents (as in

---

1. Michael Worboys, *Spreading Germs* (Cambridge: Cambridge University Press, 2006).
2. Rob Boddice, "Bestiality in a Time of Smallpox: Dr. Jenner and the 'Modern Chimera,'" in *Exploring Animal Encounters. Palgrave Studies in Animals and Literature*, edited by D. Ohrem and M. Calarco (Cham: Palgrave Macmillan, 2018), 157.

Jenner's method) or producers of antibodies to be injected into human bodies—diphtheria antitoxin being perhaps the greatest example of such an achievement.[3] Animal bodies were also tools for standardizing and testing vaccines' efficacy for the rapidly commercialized drug market.[4] Horses, guinea pigs, sheep, and other species in field farms, stables, and laboratories carried the hope for a cure to infectious diseases, tying up human and nonhuman bodies together in misery and recovery.[5]

A set of compulsory vaccination acts in 1853, 1867, and 1871 drew a sizable public critique and a movement of conscious objection, which was also prevalent in continental Europe and in the United States.[6] Opponents of human vaccination often doubted the vaccines' efficacy. They were preoccupied with notions of bodily integrity and purity or were anxious to guard their civil liberties from public health authorities.[7] However, the debates about animal inoculation experiments differed from the concerns involved in nineteenth- and early twentieth-century human vaccination. In the context of animal experimentation, the arguable pain of the disease was at the controversy's core. Inoculation research and antitoxin production tackled the Vivisection Act bureaucracy with a new kind of pain, unrelated to tissue damage. A gentle touch and a prick of a needle were not what British regulators envisioned when they drafted the Act, despite being aware of the emerging research field. Unlike the physiological experiments, which entailed preparing, strapping, cutting, and sometimes attending to the

---

3. Arthur S. MacNalty, "Emil von Behring," *British Medical Journal* 1, no. 4863 (March 20, 1954): 668–70, 669.

4. Christoph Gradmann, "Locating Therapeutic Vaccines in Nineteenth-Century History," *Science in Context* 21, no. 2 (June 2008): 145–60.

5. About the emergence of "one health," see Abigail Woods et al., "Introduction: Centring Animals Within Medical History," in *Animals and the Shaping of Modern Medicine: One Health and Its Histories*, edited by Abigail Woods et al., Medicine and Biomedical Sciences in Modern History (London: Palgrave Macmillan, 2018), 1–26.

6. For more about the history of public vaccination in the UK, see Dorothy Porter and Roy Porter, "The Politics of Prevention: Anti-Vaccinationism and Public Health in Nineteenth-Century England," *Medical History* 32, no. 3 (July 1988): 231–52; Deborah Brunton, *The Politics of Vaccination: Practice and Policy in England, Wales, Ireland, and Scotland, 1800–1874* (Rochester, NY: University of Rochester Press, 2013); E. P. Hennock, "Vaccination Policy against Smallpox, 1835–1914: A Comparison of England with Prussia and Imperial Germany," *Social History of Medicine* 11, no. 1 (April 1, 1998): 49–71.

7. Nadja Durbach, *Bodily Matters: The Anti-Vaccination Movement in England, 1853–1907* (Durham: Duke University Press Books, 2004), 4; Michael Willrich, *Pox: An American History* (New York: Penguin Press, 2011), 285. For more about the association between antivivisection and antivaccination movements in the United States, see Lederer, *Subjected to Science*, 41; Martin Fichman, "Alfred Russel Wallace and Anti-Vaccinationism in Late Victorian Cultural Context, 1870–1907," in *Natural Selection and Beyond: The Intellectual Legacy of Alfred Russel Wallace*, edited by Charles H. Smith and George Beccaloni (Oxford: Oxford University Press, 2010).

wounded animal, human interaction with the animal body in inoculation operations was often as quick as the penetration of a needle into the skin. Often, no
immediate consequences were to be seen. Symptoms appeared only gradually,
and sometimes they did not develop at all.

Moreover, even when a disease developed following inoculation, the measure
of pain involved was not easily agreed upon. Francis Sibson argued in front of
the Royal Commission that "the process of a disease is not one of constant pain,
but very much the reverse in most instances." Sibson went on to describe how,
as a physician strolling around the hospital wards, he rarely saw "any trace of
suffering on the face of any patient, particularly of a fever patient; because, so to
speak, there is an anaesthetic withing them in the very poison that saturates their
blood and solid tissues and nerves; therefore they are lying there scarcely more
than just conscious. Therefore I would say that that which applies to mankind
applies also to animals; and that they are anaesthetized by the very malady that
is produced, as a rule."[8]

In addition to the claim that pain induced by malady was milder than the
one involving direct tissue damage, research-induced disease, although manufactured by design, was often regarded as a natural occurrence and therefore
unbounded by law. Moreover, the research outcome was to benefit diseased animals, "as in the case of sheep-pox and cattle plague, but perhaps oftenest in the
common interest of both."[9]

The first Royal Commission on Vivisection divided experiments into three
categories: operations, the administration of poisonous or dangerous drugs, and
"the production of disease:—for the purpose of observing its progress, and dis-
covering the means of preventing, mitigating, or curing the effects of the same
or similar diseases in men or animals." The commission, however, failed to esti-
mate the potential of disease research, concluding that "It consists in subjecting
a comparatively very small number of animals to diseases not generally involving severe pain."[10] The commission thus did not address the specific needs of
inoculation research in its legislation recommendation. Similarly, the Vivisection
Act and Home Office implementation policies were tuned to basic physiological
research, as exemplified in the requirement to use anesthetics or to destroy animals before they recover from the tranquilizing effects—both demands unsuit-
able to most inoculation research and antitoxin production.

---

8. Royal Commission Report, 234.
9. Royal Commission Report, xiv.
10. Royal Commission Report, xiii, xv. This claim came up also in the debate about rabies
research; see Neil Pemberton and Michael Worboys, *Mad Dogs and Englishmen: Rabies in Britain,
1830–2000* (Basingstoke: Palgrave Macmillan, 2013).

Pathologists argued before the Home Office that the Act did not suit their research. They claimed that inoculation research was not an experiment per se, and at any rate, was not painful and therefore should not be governed by the Act. Hesitant, unsure, yet content to lighten its responsibilities, the Home Office asked for the help of the Crown Law Office in interpreting the Act and its relevance for disease research. The historian E. M. T. Tansey examines five out of these six opinions, showing how they manifested pressures on the emerging pharmaceutical industry.[11] Tansey sees in these opinions an expression of ambivalent attitudes toward science, but they were more about the struggle of bureaucracy with the inscrutability of pain. The following will tell the stories behind the cases while focusing on pain and experiments in the complex, sometimes contradictory, legal opinions about inoculation research and the Vivisection Act.

The Greenfield Case: Inoculation Operations Are Experiments

In November 1879, inspector George Busk received an inquiry from William Smith Greenfield (1846–1919), the superintendent of the Brown Animal Sanatory Institution. The institution, founded in 1871 by the University of London, was a center for clinical services as well as a veterinary research laboratory focusing on physiology and comparative pathology. It was the first research institution dedicated to the study of animal pathology in Britain and an early version of the more celebrated Pasteur Institute in Paris and Koch Institute in Berlin.[12] The first superintendent of the Brown Institution was the physiologist John Burden-Sanderson, whose interest in cattle plague was fostered by the great outbreak of 1865. Greenfield, Burdon-Sanderson's successor, was trained in medicine at University College London. He held the appointments of physician to the Royal Infirmary for Women and Children, and to the Royal Hospital for Diseases of the Chest, and was a physician at the department for disease of the throat at St. Thomas's Hospital, where he also taught anatomy and pathology. In 1878, he was appointed professor of pathology at the institution.

---

11. E. M. Tansey, "The Wellcome Physiological Research Laboratories 1894–1904: The Home Office, Pharmaceutical Firms, and Animal Experiments," *Medical History* 33, no. 1 (January 1989): 6. Tansey mentions five out of six opinions and neglects Roy's Case in 1882, and provides a concise summary of the ruling, with a focus on the question of production.

12. Lise Wilkinson, *Animals and Disease: An Introduction to the History of Comparative Medicine* (Cambridge: Cambridge University Press, 2005), 166; R. J. M. Franklin, "The Brown Animal Sanatory Institution—Historical Lessons for the Present?," *Veterinary Journal* 159, no. 3 (May 2000): 232.

In his letter to inspector Busk, Greenfield asked whether it was necessary for him to renew his Certificate A for his experiments with inoculation. Certificate A, which allowed dispensing with the use of anesthesia, specified: "We hereby clarify that, in our opinion, insensibility in the animal on which such experiment may be performed cannot be produced by anaesthetics without necessarily frustrating the object of such experiment." In the certificate form, signed by James Risdon Bennett, president of the Royal College of Physicians, and John Burdon Sanderson, professor of physiology at University College London, Greenfield described the proposed experiments in very general terms, as "relating to epizootic disease to be conducted by inoculation or by otherwise exposing animals to contagion at the Brown Institution."[13] Given that anesthesia was the major reassuring device for those who were concerned with animal pain, Certificate A for experimenting without anesthesia was the most problematic for public opinion. As later explained by an antivivisection writer, "Certificate A is granted specially to exempt the experimenters from the use of anaesthetics, and therefore, all these extra experiments *were carried out on animals sensitive to pain*."[14]

At the time he composed the letter, Greenfield was researching septicemia (blood infection) and anthrax, and arguably anticipated Louis Pasteur in preparing an effective vaccine against the disease. Greenfield attenuated the disease agents by a successive cultivation, mitigating the fatality of the bacterium *Bacillus anthracis* by passing it through the bodies of guinea pigs or mice, then inoculating cows with the rodents' blood or spleen.[15]

Greenfield claimed in the letter that his planned experiments were "not in any sense painful" and that "in a large number of cases they do not produce any effects. In other cases, the suffering to the animals in consequence of such inoculation is practically nil." Public opinion provided another kind of argument against burdening inoculation experiments with a certificate: "It appears scarcely desirable to swell the statistics of supposed painful experiments by the record of such inoculation in future." According to this reason, including inoculation

13. Copy of Certificate A, February 17, 1879, HO 144/35/81756.

14. "State-Recognised Cruelty," *Cheltenham Chronicle* 4301, July 2, 1892, 1, 19th Century British Library Newspapers Database.

15. Wilkinson, *Animals and Disease*, 174; W. D. Tigertt, "Anthrax. William Smith Greenfield, M.D., F.R.C.P., Professor Superintendent, the Brown Animal Sanatory Institution (1878–81). Concerning the Priority Due to Him for the Production of the First Vaccine against Anthrax," *Journal of Hygiene* 85, no. 3 (1908): 415–20; W. S. Greenfield and John Scott Burdon-Sanderson, "IX. Preliminary Note on Some Points in the Pathology of Anthrax, with Especial Reference to the Modification of the Properties of the Bacillus Anthracis by Cultivation, and to the Protective Influence of Inoculation with a Modified Virus," *Proceedings of the Royal Society of London* 30, no. 200–205 (January 1, 1880): 557–60.

research in the Home Office's statistics of experiments on living animals would be deceptive since inoculation experiments required many animals while "often none, and on other cases very few are affected." Greenfield thus inquired whether "the exposure of animals to infection is considered a painful experiment under the Act."[16]

Inspector Busk briefed Home Secretary Cross about Greenfield's inquiry. Busk divided the inoculation operation into two phases, and claimed that no question arose in relation to the first stage, the "mere operation of inoculation" since "it cannot be regarded in itself as 'calculated to give pain,'" and therefore should not be restricted by the Act. However, the "occasional after consequences in the communication of the disease and its concomitant suffering" may bring the experiment within the provisions of the Act. The same logic, Busk claimed, was to be applied to experiments by exposure to infection by other means than needle puncture.

Busk was inclined to leave inoculation research ungoverned by the Act. He reasoned that in many cases no effect was produced by the inoculation, and in the cases in which disease did develop, the infected animal was "no worse off than one which has contracted the disease in the usual way and suffers no more; in fact, it might be regarded as so far better off that the surrounding condition will be more favourable and the treatment it meets with probably more immediately beneficial."[17]

The promising research carried on at the Brown Institution was another reason to allow the operations to go unsupervised. Busk emphasized that the research was "undoubtedly one of extreme and even urgent public importance" and that "the fewer restrictions imposed upon it and the fewer interruptions to which it may be liable, the better." He therefore believed that experiments such as those conducted by Greenfield should be "left as unaffected as the law will allow."

Additionally, there were potential incoherencies with imposing the Act on inoculation in research facilities. It would make no sense, Busk contended, to allow owners of stocks to inoculate or expose their animals to infection as means of preventing more severe outbreaks, while restricting a similar practice "which is so ably and carefully conducted at the Brown Institute; and which has solely for its object the prevention, cure or mitigation, of diseases of the greatest

16. W. S. Greenfield to George Busk, November 17, 1879, Cruelty to Animals Act 1876: Case to the Opinion of the Attorney and Solicitor General, HO 144/35/81756.
17. Busk to Secretary of State, November 19, 1879, Cruelty to Animals Act 1876: Case to the Opinion of the Attorney and Solicitor General, HO 144/35/81756.

economical importance, and which experiments moreover seem to be attended with scarcely any appreciable suffering."[18]

Legal assistant undersecretary Lushington forwarded the case, marked "pressing," to the Law Officers, probably following the instruction of Home Secretary Cross.[19] The Law Officers were asked whether inoculation experiments and those experiments that are constituted by "exposure of animals to injection" came within the scope of the Act and required licenses. The case listed the arguments in favor of excluding these experiments from the Act, including the concern about Certificate A bad public reputation: "in the public mind such a certificate is thought to imply experiments of a very different and much more painful character, and therefore the issue of such certificate in a case of inoculation ought not to take place unless it is made absolutely imperative by the statute."[20]

The Home Office brief to the Law officers also laid out the argument in favor of requiring a license and a certificate from experimenters. The document compared inoculation followed by a disease with the pains after a surgical operation under anesthetics: "The inoculation is not less painful than the surgical experiment, which being done under anaesthetic, is absolutely painless, whilst the after effects of the inoculation disease, perhaps mortal disease—may for the present purposes be assumed to be more painful than the after effects of the surgical experiment." In other words, under the assumption that the object of the Act is to protect animals from pain, if surgical operations required a certificate that allows keeping an animal alive after experiments, a license and a special certificate should also be required for the potentially more painful inoculation experiments.

The case was handed to Attorney General John Holker and Solicitor General Harding Giffard. Their opinion, short and decisive, concluded that inoculation operations were under the Act: "We think both sets of proceedings are experiments upon living animals within the meaning of the Act and therefore require the licenses. It may be that the subject prominently in the mind of the legislature was something in the nature of vivisection but the language of the statute is too wide to permit of it being limited to experiments of that class nor do we see any part of the Act upon which such a limitation could be supported."[21] In

18. Busk to Secretary of State, November 19, 1879, Cruelty to Animals Act 1876: Case to the Opinion of the Attorney and Solicitor General, HO 144/35/81756.

19. Lushington to the Solicitor of the Treasury, November 23, 1879, HO 156/1, 306.

20. Opinion by John Holker and Harding Giffard, December 3, 1879, L.O.O. 584-1, HO 144/35/81756.

21. Opinion by John Holker and Harding Giffard.

other words, the Law Officers acknowledged that if interpretation of the Act was bound by the intention of the legislature, the term 'experiment' was restricted to vivisection in its literal meaning of cutting into the flesh of a living body. But the Act was three years old and already partially outdated as inoculation experiments were proliferating.

Inoculation operations were decided by the Greenfield Case to be experiments like any other. After leaving the Brown Institution in 1881, Greenfield lamented that his experiments were underfunded and stalled "in the face of all the difficulties interposed by law; whilst M. Pasteur is encouraged and abundantly supplied with means by the liberality of the French Government."[22]

Busk, who opposed the inclusion of inoculation operations from the beginning, presented inoculation as a painless procedure in his annual to the Parliament despite the Law Officers' opinion. In his 1880 report, Busk mentioned thirty experiments on guinea pigs and mice, where "disease appears to have ensued, which, during the brief period the animals survived, may have caused a slight suffering."[23] In Busk's next annual report, he referred to inoculation experiments as "no more painful than the prick of a lancet or a needle."[24] After about a decade in which the expression "prick of a needle" was used in the annual reports, the *Zoophilist* published a letter from Francis Power Cobbe under this title. Cobbe argued that the way the inspectors described inoculation experiments in their annual reports was "an imposition on the British public."[25]

The Roy Case: Redefining Experiments

The opinion in the Greenfield Case made clear that research that involved inoculation came under the Act. Yet two years later, Greenfield's successor in directing the Brown Institution provoked the same debate. Charles Smart Roy (1854–97) was born in Scotland, and received his medical degree from the University

---

22. W. S. Greenfield, "Inaugural Address on Pathology, Past and Present," *British Medical Journal* 2 (1881): 733. Also in Wilkinson, *Animals and Disease*, 174.

23. Experiments on Living Animals: Copy of Report from the Inspectors Showing the Number of Experiments Performed on Living Animals During the Year 1880, under Licences Granted Under the Act 39 & 40 Vict. c. 77, Distinguishing Painless Experiments from Painful Experiments, Parliamentary Papers, 1881 (298), 3.

24. Experiments on Living Animals: Copy of Report from the Inspectors Showing the Number of Experiments Performed on Living Animals During the Year 1881, under Licences Granted Under the Act 39 & 40 Vict. c. 77, Distinguishing Painless Experiments from Painful Experiments, Parliamentary Papers, 1882 (165), 3.

25. "The Prick of a Needle," *Zoophilist* 11, no. 4 (1891): 62

of Edinburgh. After serving as a resident physician at the Edinburgh Royal Infirmary, he moved to the Brown Institution to take up research in physiology and pathology. During a leave, he practiced as an assistant to Friedrich Goltz in the Strasbourg Physiological Institute and conducted research on the kidney, heart, and spleen in various laboratories. He was appointed to direct the Brown Institution in July 1881.[26]

In December 1881, Roy applied to the Home Office for a certificate B, which would allow him to keep dogs and other animals alive after inoculating them with distemper, a virus affecting respiratory, gastrointestinal, and nervous systems. The operation consisted "simply" of piercing "slightly with a needle . . . a minute portion of the skin." The effects of the operation "if any, will be to cause a disease which all or nearly all dogs suffer at one period of their life and which seldom or never attacks the same animal twice."[27] Lushington deferred his decision maintaining that the application was incomplete, and asked Roy to duly submit a signed certificate.[28]

In his next letter, Roy explained that he planned to dispense with the use of anesthesia but "shall at once have recourse to them, and in all other respects comply with the requirements of the Act in case reason should arise supposing that pain is inflicted."[29] Instead of submitting the standard forms for his operation, Roy attached to his note a support letter (which he called a "certificate") signed by William Jenner, the president of the Royal College of Physicians. Jenner certified that Roy's experiment "will not be attended with the infliction of any appreciable amount of pain on the animals used, excepting in so far as they may eventually be productive of disease." Jenner also added that although anesthesia might not frustrate the object of the experiment it had "no amount of benefit to the animals" and hence its use was unnecessary.[30]

Inspector Busk required Roy to submit the usual forms replacing the letter of support written by Jenner. At that point, he did not see anything unusual in the application and added that Roy would require certificates A (dispensing with anesthetics) and E (experiments on dogs when dispensing with anesthesia), or at least the latter.[31] Roy was quick to compose an indignant reply. The employment

---

26. "Charles Smart Roy, M.A., M.D., F.R.S., Professor of Pathology In The University of Cambridge," *British Medical Journal* 2, no. 1919 (1897): 1031–32.

27. Roy to Harcourt, December 13, 1881, HO 144/67/98256.

28. Lushington to Roy, December 14, 1881, HO 156/1, 406.

29. Roy to Harcourt, December 21, 1881, Cruelty to Animals Act 1876 Case, HO 144/67/98256.

30. William Jenner to Vernon Harcourt, December 21, 1881, in Cruelty to Animals Act 1876 Case, HO 144/67/98256.

31. Busk to Secretary of State, December 27, 1881, in Cruelty to Animals Act 1876 Case HO 144/67/98256.

of anesthesia, he argued, will oblige him to subject the animals to "unnecessary illness and misery." He offered "one or two facts which may perhaps lead you to modify your decision on this subject." First, the inoculation itself—as certified by Jenner—caused no pain "and to use chloroform while making the few scratches required is to expose the animal unnecessarily to the sickness and misery which always follows in dogs after the anaesthetization under chloroform." He explained that it was inappropriate to ask the president of the College of Physicians and a professor to sign the ordinary certificate in which they would have to confirm that "the use of an anaesthetic would necessarily frustrate the object of the experiments," because that would be a false assertion. The use of anesthetic, Roy claimed, "would only subject the animals mercilessly to cruel misery which I am extremely anxious to avoid."[32]

In addition to arguing that anesthesia would be painful and unnecessary, Roy claimed that artificially causing a mild form of distemper was "really an advantage for any dog" since dogs would predictably be infected with the disease at some time in their lives, while "in this institute (where I have all the appliances for the proper nursing of sick dogs with well trained and kind hearted servants) they would be better taken care of during their illness than they possibly could be elsewhere."[33]

But there was another reason why Roy resisted submitting the usual forms required by the Act, revealing that he was concerned about his reputation as much as about the welfare of his dogs. Roy calculated that for his proposed experiments, he would need at least three different certificates: A, B, and E. This would increase the number of certificates in the Home Office records (with which "I have nothing to do"), but he was mostly concerned that submitting these certificates would draw unwanted attention to the Brown Institution. Roy predicted that some people "would naturally conclude that cruel and severe experiments were being performed here that these might then arise a popular outcry against this Institution which would be harmful for its interests," and expressed his hope that Home Secretary William Harcourt "will lay much weight upon this latter reason."

Roy therefore argued that Jenner's statement according to which his experiments were painless should satisfy the home secretary, and reemphasized that any other requirement would "necessitate the subjection of the animals employed to unnecessary cruelty thereby violating the spirit if not the letter of

---

32. Roy to Harcourt, December 27, 1881, in Cruelty to Animals Act 1876 Case, HO 144/67/98256.
33. Roy to Harcourt, December 27, 1881, in Cruelty to Animals Act 1876 Case, HO 144/67/98256.

the Act."[34] Here Roy made a legal argument. His understanding of animal bodies and pain made him argue for a privileged perspective over the "spirit" of the Act. Thus, while the Home Office administration and Law Officers were deciding whether inoculation experiments were painful or not, experimenters took pains in advancing their interpretation of the Act and its intention.

Roy had Busk on his side. The inspector was convinced that administering anesthesia for inoculation caused more suffering than the operation itself. He also agreed that the Act—which allowed dispensing with anesthesia only when its use would frustrate the objectives of the experiment—did not provide a suitable solution considering the nature of inoculation experiments. In his mind, it was "only reasonable to expect that no scientist would be willing to sign such a doubtful statement." Busk added that Roy's arguments were novel and had not been invoked before by any of the applicants for certificates. Busk acknowledged that the Law Officers' opinion in the Greenfield Case was still binding but opined that "considering the painless nature and so far as the animal itself is concerned the beneficial effect of the proposed experiment they cannot justly be regarded as 'experiments calculated to give pain.'" The Home Office should therefore not require Roy to obtain a certificate to carry out his "useful object."[35]

Lushington was baffled. On the one hand, Roy's arguments against requiring certificates A and B for inoculations were sound. On the other hand, he was bound by the law and its interpretation by the Law Officers, and he thought it was impossible for the home secretary to sign certificates that were "not based on the ground which the Act specifies as their justification." Lushington thought it was best if Roy carried out his experiments without a license or certificate. But he also believed that this was "a question for a court of law," and that the home secretary "must not be understood to express any opinion concerning it."[36] Home Secretary Harcourt forwarded the case to the Law Officers, with a copy to the Medical Research Association. The question was phrased as such: "Whether they concur with the opinion of the former Law Officers that inoculation for the purpose of research is an experiment calculated to give pain within the meaning of the statute."[37]

There are several possible reasons why Harcourt would ask for a second legal opinion on a subject that had been decided as recent as two years earlier. It might

---

34. Roy to Harcourt, December 27, 1881, in Cruelty to Animals Act 1876 Case, HO 144/67/98256.
35. Busk to Secretary of State, December 28, 1881, HO 144/67/98256.
36. Minute by Godfrey Lushington, December 29, 1881, HO 144/67/98256.
37. Lushington to the Solicitor of the Treasury, December 30, 1881, *in* Cruelty to Animals Case, HO 144/67/98256.

have reflected distrust in the judgment of his predecessor at the Home Office. It might have been an attempt to ease the increasing burden on the Home Office, as several social regulations in the 1870s made new demands on officials and loaded the office with paperwork. In the early 1880s the Home Office reacted to pressure by trying to have some of its responsibilities removed.[38] It might also have expressed a hope that a new opinion would spare Harcourt a confrontation with the scientists. Harcourt was an acquaintance of Joseph Hooker and Thomas Huxley and sympathized with the cause of vivisection; he was later in charge of creating the arrangement that allowed the Association for the Advancement of Medicine by Research (AAMR) to participate in the evaluation of license applications.[39] And maybe it was a genuine legal concern: Harcourt's decision might have responded to the arguments articulated by Roy, which had not been addressed in the Greenfield Case.

The case was handed to Henry James, the newly appointed attorney general and Farrer Herschell, the solicitor general. Lushington received a first, informal, response from Attorney General James, stating that he did not think inoculation came under the Act.[40] Harcourt was too quick in passing the information along to Roy, probably in an attempt to allay the pressures coming from Roy.[41] In a few days, the interpretation that inoculation was not under the Act had been unexplainably overturned. In the official opinion delivered to the Home Office on January 11, 1882, Law Officers James and Herschell expressed their reservations about the opinion in Greenfield's case, yet declined to change their predecessors' interpretation of the Act since the question involved "grave doubts":

> If this case had come before us in the first instance we should have expressed great doubts whether the artificial production of disease by the act of inoculation (in the absence of pain in the act itself) would be an experiment calculated to give pain . . .
>
> But as we feel that the question involves grave doubts and that either view of it may be supported by very substantial arguments we are not disposed to express so confident an opinion in opposition to that of the late Law Officers as would justify the Home Secretary in authorizing the Act of inoculation to take place without a certificate.[42]

38. Pellew, *The Home Office*, 39.

39. French, *Antivivisection and Medical Science*, 184, 206. French, however, argued that Harcourt's arrival at the Home Office "worsened rather than improved" the situation from the perspective of experimental medicine. French, *Antivivisection and Medical Science*, 188.

40. Henry James to Lushington, January 6, 1882, HO 144/67/98256.

41. As an example of the pressure, see Roy to Harcourt, December (should be January) 2, 1882, HO 144/67/98256; Lushington to Roy, January 9, 1882, HO 156/1, 425.

42. Opinion, L.O.O. 584/2, January 11, 1882, in Cruelty to Animals Act Case, HO 144/67/98256.

James and Herschell thus unenthusiastically reiterated the opinion in the Greenfield Case that inoculation experiments were governed by the Act. The following part of their opinion however explicitly expanded the time range of the inoculation experiment, thus solving what was presented by Roy to be an irrational requirement to anesthetize animals for the moment of injection: "The experiment must be regarded as extending during the whole time over which the disease is being produced and lasts and therefore it may safely be certified that insensibility by anaesthetic cannot be produced without necessarily frustrating the object of the experiment." In other words, the Law Officers refuted the claim that Certificate A was inappropriate for inoculation, by redefining the inoculation experiment as starting with the needle and continuing through the disease period. Seen this way, the authorized persons should not refrain from signing Certificate A for inoculation.

The definition of an experiment advocated by the Law Officers was an intervention that unintentionally redefined the fate of other kinds of experiments such as brain experiments that were discussed in the previous chapter. At the same time, as predicted by Roy, it increased the number of experiments listed as operations without anesthesia and alarmed the antivivisection societies.

Following the legal opinion, Roy submitted a standard form of Certificate E (experiments on dogs without anesthesia) approved by William Jenner and John Burdon Sanderson in April 1882. The description of the experiments was "inoculation of the mitigated virus of distemper with the view of finding a means of producing inoculation where-by the mortality caused by distemper in dogs may be diminished or done away with." The reasons specified for the use of dogs over other animals were that "dogs are the usual victims of distemper," and because an investigation of the effects of inoculation with distemper virus on other animals "would be of little value."[43]

Indeed, being listed as a vivisector of dogs was socially costly. Roy's concerns about his reputation being damaged by the publicity of the certificates became real when in October 1882 the Victoria Street Society published a pamphlet under the title "The Veracity of the Parliamentary Returns as Illustrated by the Case of Dr. Roy."[44] The pamphlet accused the Home Office inspectors of covering up Roy's uncertified and therefore illegal experiments that he conducted during 1880 and 1881. Lord Shaftesbury, who was active in the establishment

---

43. Roy Charles, Certificate E, April 3, 1882, HO 144/67/98256. Liddell informed Roy about the decision. Liddell to Roy, January 14, 1882, HO 156/1, 428.

44. Society for the Protection of Animals from Vivisection, *The Veracity of the Parliamentary Returns as Illustrated by the Case of Dr. Roy*, n.d., HO 144/67/98256.

SURGICAL.

Section 25.

Dog Holder. Improved from Bernard's Model and suited for both large and small animals.

Dog Holder.

Roy's Dog Holder. It securely holds the dog's head in any position without causing pain and it can be used with dogs of all sizes.

Rat Holder. This is for the same use as the Dog Holder; but it is differently constructed and suitable for a rat; it has also a tube for applying the anæsthetic.

FIGURE 4 / "Roy's Dog Holder," Cambridge Scientific Instrument Co., *A Descriptive List of Instruments Manufactured and Sold by the Cambridge Scientific Instrument Company* (Cambridge: 1891), 103. Whipple Museum of the History of Science, University of Cambridge, CSI.C1. Roy "invented many ingenious pieces of apparatus for physiological purposes," which "will continue to be called by his name," in "Charles Smart Roy," *British Medical Journal* 2, no. 1919 (1897): 32.

of the Victoria Street Society, demanded the home secretary's response to the claims.[45]

Lushington drafted a reply based on a document prepared by Busk, in which he confirmed that Roy held a license in 1880 and 1881 but no special certificate, adding that Roy indeed experimented on dogs and cats "openly," but those were different experiments than the inoculation experiments he applied for in late 1881.[46] The home secretary made his investigation "and satisfied himself that the condition as to anaesthetics was fully complied with." Lushington added that since 1882 Roy had held the right certificates. He concluded that the charges against Roy were unjustified, and "no less unjustifiable is the charge that this license or the certificates have been of a cruel character."[47]

Agitated, Roy read Victoria Street's pamphlet aloud in a meeting of the Executive Committee of the AAMR. The Committee resolved to "request the secretary to prepare a statement vindicating Dr. Roy's character."[48] Roy was thankful for the resolution the AAMR had published, but he was still carefully watched by critics of vivisection.[49] The *North-Eastern Daily Gazette*, for example, warned its readers against having their dogs and cats taken by research institutions and in particular "Brown Institute for Dogs, where Dr. Roy, a young Scotchman, who had gained some notoriety as a vivisector of an unusually cold-blooded type, presides."[50]

The AAMR on Pain

The AAMR pushed against the two legal opinions and their evaluation of pain. Its standpoint was that inoculation experiments were relatively harmless. In June 1882, shortly after the above correspondence between Roy and the Home Office, the AAMR published the "Memorandum of Facts and Considerations Relating to the Practice of Scientific Experiments on Living Animals Commonly Called Vivisection." While the AAMR's main line of argumentation emphasized the importance of physiological research for the advancement of knowledge and

---

45. Society for the Protection of Animals from Vivisection, *The Veracity of the Parliamentary Returns as Illustrated by the Case of Dr. Roy*, 3; about Shaftesbury, see Edwin Hodder, *The Life and Work of the 7th Earl of Shaftesbury* (London: Cassell & Company, 1887), 696.

46. Letter draft from Busk to the Secretary of State, October 30, 1882, HO 144/67/98256.

47. Letter draft from Lushington to Harcourt, November 1, 1882, HO 144/67/98256.

48. Executive Committee Seventh Meeting, November 7, 1882, WA, MS 5310, 21.

49. Roy to the AAMRR, January 28, 1883, Minutes of the Executive Committee: Ninth Meeting, January 30, 1883, WA, MS/5310, 23.

50. "Honest Poverty," *North-Eastern Daily Gazette*, June 24, 1884, British Newspaper Archive.

the betterment of medicine, it made continual efforts to downplay the suffering involved in the practice of vivisection.[51]

The AAMR classified experiments into five categories based on the quality and quantity of pain they induced: first were those which were entirely unaccompanied by pain and could be performed "either upon animals or upon man himself." In this category were various experiments on vision, taste, smell, and touch as well as bodily heat, pulse, and respiration. In the second were observations of tissues and organs of dead animals, including experiments on the action of the heart "which in cold-blooded creatures continues long after their death." The third category was that of anesthetized animals. These experiments were "carried out without any pain or even discomfort to the animals," afterward they were "deprived from life in probably the most painless manner possible."[52]

A fourth category included experiments that allowed an animal to recover from the operation. In these cases, the "severest pain" of the operation was alleviated with anesthetic, and thereby remained only a "subsequent suffering . . . quite insignificant," such as of a healing wound or inflammation, colic, or fever. The memorandum then referred to acupuncture or inoculating operations and, isolating the moment of the prick of the needle, claimed that the initial pain involved in many of these experiments was often "so trifling that it would be unreasonable to give an anaesthetic." It would be "unreasonable to give a rabbit chloroform for such operations as bleeding, vaccination . . . for which no human being would take it." The fifth and last category of experiments was disease research. The pain involved with disease was however "more justly described as discomfort than as torture." The subsequent test of certain drugs or treatments such as inoculation, produced suffering which was "less than the familiar effects of corresponding remedies in human beings."[53] The AAMR acknowledged the existence of pain only in the two categories of diseased animals and animals recovering from operations, and even there, argued pain to be mild or manageable by anesthesia.

The status of inoculation experiments, the pain they produced, and their inclusion under the Act was constantly challenged, and disapproval was voiced even from within the Home Office. During a parliamentary debate in April 1883, Home Secretary Harcourt expressed his view, that "there is one class

---

51. AAMR, "Memorandum of Facts and Considerations Relating to the Practice of Scientific Experiments on Living Animals, Commonly Called Vivisection," June 1882, WA, SA/RDS/A4, 5.

52. AAMR, "Memorandum of Facts and Considerations Relating to the Practice of Scientific Experiments on Living Animals," 6.

53. AAMR, "Memorandum of Facts and Considerations Relating to the Practice of Scientific Experiments on Living Animals," 7.

of experiments which, by a misnomer, are called vivisection, and which are performed without anaesthetics—inoculation experiments . . . does anybody con-
demn inoculation for the purpose of seeing the effect of contagious diseases on
the lower animals? At least, I do not; and I do not think they can be, or ought to
be, condemned."[54] A decade later Home Secretary Herbert Asquith explained that
inoculation experiments were allowed to be done without anesthesia "because
they are perfectly painless."[55]

Whatever the Home Office administrators thought about inoculation experiments and the pain they produced, they followed the Law Officers' opinions
in imposing the Act's restrictions. In November 1885, Edward Schäfer, then a
professor at University College London, refused to sign an application for certificate A for inoculation experiments on the grounds that this would be a false
assertion. Busk agreed, "this is quite true," but explained the policy to Schäfer,
stating that the Law Officers decided to include it under the Act "apparently on
the ground that the mere operation is only the beginning" of what may be "a long
and painful malady."[56]

The Inoculation (Pain) Condition

Lushington was troubled by the challenges posed by diseased animals. He was
disturbed by the Act's shortcomings in dealing with inoculation experiments,
and did not believe licenses and certificates were adequate to fulfill the Home
Office's responsibility under the Act. He asked Home Office inspector John Eric
Erichsen to gather data about the various kinds of experiments and they brain-
stormed about certificates A and B.[57] In January 1887, Erichsen composed a special report on inoculation experiments. Erichsen suggested attaching a condition
to the license of any person performing inoculation experiments with Certificate
A. The condition would require the licensee to put the animal to death under or
by anesthetics when the main result of the experiment had been attained and
at any time during the experiment if continuance of life would necessarily be

---

54. 277 Parl. Deb. (3rd ser.) (1883) 1440.
55. 17 Parl. Deb. (4th ser.) (1893) 344.
56. Busk to Schäfer, November 16, 1885, WA, PP/ESS/B27/1-15. The instructions were sometime
confusing, as in the next example: "certificate A covers simple cutaneous inoculation only, but, if you
contemplate the introduction of a virus into the deeper parts of the body or if you intend to use any
instruments other than a simple hypodermic syringe" it will be necessary to ask for certificate B. A letter on behalf of the Home Secretary to Louis Cobbett, October 31, 1892, HO 156/7, 279.
57. Memo by Lushington, HO 144/17/44209X; Erichsen to Lushington, December 9, 1886, HO
144/17/44209X.

attended by prolonged and severe pain. Based on a conversation with James Paget, chairman of the AAMR, Erichsen estimated that there would be no objection to such an additional condition—even by those engaged in inoculation experiments at the Brown Institute.[58]

Home Secretary Henry Matthews accepted Erichsen's recommendations and introduced the "inoculation condition," also referred to as the "pain condition." Lushington, now the undersecretary, issued a circular to the inspectors explaining that since "the duration of the experiment must be taken to continue from the first puncture until the animal dies, is killed, or recovers" a condition will be attached to the license: "If severe pain had been induced in an animal after any of the said experiments has been performed . . . and if the main result of the experiment has been attained, it is a condition of this license that the animal be immediately killed under anaesthetics."[59]

All licensees received a notification about the new condition, and those involved in inoculation experiments were asked to forward their license back to the Home Office to be updated.[60]

In 1889, two years after its introduction, the inoculation (pain) condition was modified and the degree of pain requiring the killing of a suffering animal changed: if an animal "is found to be in pain which is either considerable in amount or is likely to endure, and if the main result of the experiment had been attained, the animal shall be immediately killed under anaesthetic."[61] The term "severe pain" in the original condition was replaced by the alternative markers of considerable or enduring pain, meaning that more cases were then potentially influenced by the directive.

In Erichsen's annual reports for 1886, 1887, and 1888, he contended that most experiments that were allowed to dispense with the use of anesthetics (under Certificate A) were comprised of a "simple puncture" or "simple inoculation experiments and were consequently painless." Erichsen was undoubtedly aware that these reports were closely examined by antivivisectionists, and described the inoculation condition as requiring experimenters to destroy

---

58. John Eric Erichsen, "Inoculation Experiments," January 10, 1887, HO 144/17/44209X.

59. Letter Draft by Lushington, January 18, 1887, HO 156/2, 620; Memorandum by Dr. Thane, [1906?], HO 45/10521/138422.

60. A letter template by Godfrey Lushington, January 28, 1887, HO 144/17/44209X. See also the testimony of the Home Office clerk W. P. Byrne, in Royal Commission on Vivisection, *Appendix to First Report*, 4.

61. Memorandum by Dr. Thane [1906?], HO 45/10521/138422.

animals under anesthetics in the event, which "very rarely happens," of pain having been developed.[62]

Interestingly, as Erichsen anticipated, there is no evidence of meaningful resistance by experimenters to the inoculation condition, unlike to other restricting conditions, such as the one limiting experiments' duration. One reason for this was revealed in a Home Office memorandum sent about two decades later to the secretary of the Royal Commission on Vivisection asking the commission to deliberate upon changing the pain condition so that experimenters would be required to destroy suffering diseased animals regardless of whether the main result of the experiment had been attained. The home secretary expected that such a change "would be followed by a good deal of controversy with experimenters as to whether animals in particular cases were in pain or not."[63] In other words, it was easier for experimenters to decide whether they concluded their experiment than evaluating the suffering of diseased animals. In addition, the memorandum explained that "the reason why the condition has been framed as it now stands is that a condition containing an absolute requirement to kill the animal at once on the supervention of pain would prevent a certain class of experiments, small in number, but of a great value, being performed, or at least carried to any useful issue."[64]

Since the inoculation (pain) condition left it to the experimenters' discretion to decide whether the main result of their experiment had been attained, the commission recommended adding to all certificates, inoculation and others, an altered version of it: "That an inspector should have power to order the painless destruction of any animal which, having been the subject of any experiment, shows signs of *obvious suffering or considerable pain*, even though the object of the experiment may not have been attained" and that "in all cases in which in the opinion of the experimenter the animals is suffering *severe pain which is likely to endure* it shall be his duty to cause its painless death, even though the object of the experiment has not been attained."[65]

---

62. Experiments on Living Animals: Copy of Report from the Inspectors Showing the Number of Experiments Performed on Living Animals During the Year 1886, under Licences Granted Under the Act 39 & 40 Vict. c. 77, Distinguishing Painless Experiments from Painful Experiments, Parliamentary Papers, 1887 (136), 4; Experiments on Living Animals, Parliamentary Papers, 1888 (186), 4; Experiments on Living Animals, Parliamentary Papers, 1889 (114), 4; Experiments on Living Animals, Parliamentary Papers, 1887 (136), 4.

63. Henry Cunynghame to the Secretary of the Royal Commission on Vivisection, April 17, 1909, HO 114/2.

64. Form of "Pain" Condition Usually Attached to Licenses in Respect of Certificate A, in Henry Cunynghame to the Secretary of the Royal Commission on Vivisection, April 17, 1909, HO 114/2.

65. Royal Commission on Vivisection, *Final Report*, 64; emphasis added.

Three members of the commission (A. R. M. Lockwood, William J. Collins, and G. Wilson) dissented, arguing that the modification of the condition was incoherent: if the inspector was empowered to require the killing of an animal showing signs of obvious suffering or considerable pain even before the experimenter attained his results, there was no reason why an identical instruction would not be directed toward experimenters without the addition that the pain was "likely to endure."[66] The committee members did not, however, point to the recurring difficulty in distinguishing "considerable," "obvious," and "severe" pain. In any case, even with the new condition, if no inspector was present (which means, in most circumstances), deciding when the experiment ended and whether an animal was in pain was under the experimenters' control.

Discontent at the Administration of the Act

The establishment of the AAMR in 1882 was motivated by the hope of making the Act into a "registration act." But its members did not envision how many struggles and efforts were embedded even in the simplest act of registration. In late January 1889, Roy complained in front of the AAMR about the "delays, difficulties and inconsistencies" in the granting and renewal of licenses and certificates.[67]

A subcommittee on the matter had gathered soon after, composed of James Paget, Edward Schafer, Victor Horsley, (Clinton) Dent, and Philip Pye-Smith. After going through some letters on the subject, among which were letters by Michael Foster and Gerald Yeo, the subcommittee drafted its conclusions. As could be expected, the subcommittee urged granting the AAMR additional power over managing applications for licenses and certificates, including the exclusive authority to limit the number of experiments in certificates. As for inoculation, the subcommittee stated that "being almost painless, should not need any certificate," and restated that Certificate E, which was required for experiments on dogs and cats without anesthesia was "illegal, and that returns made under these are false; and that the Home Office ought not therefore to demand it." The AAMR also contended that "in general, the attitude of the Home Office toward those engaged in research, and the addition lately made by the Home Secretary to the Act, are not such as men of science have a right to expect."[68]

---

66. Royal Commission on Vivisection, *Final Report*, 71.

67. Minutes of the AAMR Executive Committee, Minute Book of the Association for the Advancement of Medicine by Research, January 25, 1889, WA, MS 5310, 53.

68. Minutes of the First Meeting of the Sub Committee for the Memorial to the Home Office, Minute Book of the Association for the Advancement of Medicine by Research, February 5, 1889, WA, MS/5310, 54–55.

A deputation of AAMR members then met Home Secretary Matthews to lay out their claims. Following the meeting, the latter informed that his recent policies regarding limiting the number of experiments and setting the date of expiration for certificates would be relaxed in "simple cases."[69] However, the home secretary rejected most of the AAMR's other complaints, including those related to the controversial Certificate E. Matthews was confident in adopting his predecessors' policy to require the certificate also for inoculation and brain operations. Again, the Home Office rationale was that experiments in which animals recovered from anesthesia included two stages. The second stage of observing the result of an operation "may extend over days, weeks or months and the experiment continues until the final result is obtained or the animal dies." It was not always clear whether the animal was in pain during this stage, a state that "must depend upon the nature of the experiment and the care with which the animal is treated." Therefore, "for the purpose of protecting dogs and cats" it was necessary to impose an additional Certificate E covering the second stage. If dogs and cats were "kept alive to suffer pain" and the Home Office would not have required that certificate, "great public dissatisfaction would be felt" leading, warned the Home Office, "to more restrictive legislation."[70]

Roy was irritated by what he conceived to be a submissive attitude of the AAMR. He was also probably not less upset to receive in July 1893 a circular from Edward Leigh Pemberton, legal assistant undersecretary, delivered to all licensees. The circular warned that Home Secretary Asquith was concerned about several cases in which licensees "have failed to comply with the requirements of the statute, and in excuse have pleaded their ignorance of the law ... Mr. Asquith greatly regrets these infringements ... he will be compelled to exercise the powers conferred on him by the Act, and adopt the severe course of invoking licenses and disallowing the certificates." Pemberton also clarified that Certificate A for dispensing with the use of anesthesia was necessary also in cases of "simple inoculation not involving any surgical operation."[71]

In reply, Roy composed a five-page Statement on the Administration of the Act in which he complained that the Home Office did not consult experts when deciding upon matters of scientific research, and argued that inspectors Busk and Erichsen (who were no longer on duty) informed him "that they have been unable to induce the Secretary of State to accede to certain recommendations." Roy contended that the home secretary transgressed his authority with additional restrictions on certificates, and his actions were "harmful to the interests

---

69. Lushington to the Secretary of the AAMR, May 3, 1889, HO 144/315/B7414A, 6; Memo, January 24, 1891, HO 144/315/B7414A.

70. Lushington to the Secretary of the AAMR, May 3, 1889, HO 144/315/B7414A, 8, 10.

71. Letter template by Leigh Pemberton, July 12, 1893, HO 144/315/B7414A.

of humanity." Roy also expressed his dissatisfaction with the AAMR (of which he was one of the original members) that "has failed in the objects for it was formed."[72]

The Home Office treated Roy's criticism seriously, as testified by the many drafts it prepared in reply.[73] In his formal reply, Home Secretary Asquith clarified that he had the authority to annex conditions to licenses. He also rejected the claim that the conditions that he added "originated from any scientific views of his advisers" which conflicted with other expert views, but rather claimed his decisions were based on "more general considerations of what appeared necessary for the proper administration of the Act and the satisfaction of Parliament."[74] In other words, the Home Office claimed its policies were strictly bureaucratic.

When Do Experiments End and Production Begin?

Experimenting with inoculation primarily challenged the interpretation of the concept of pain, but at the same time it led to the reexamination of the definition of experiment. It began with the question of whether disease research extended beyond the initial stage of a needle's puncture in Greenfield and Roy cases from 1879 and 1882 and became complicated in the following years with the use of inoculation as a preventive method, and even more with the production and testing of antitoxins.

Swine fever outbreaks were not the severest epizootic threats in the late nineteenth century, but they drew considerable attention. From 1878 to 1892, the main measures employed to suppress the epidemic were the slaughter of diseased animals and those that had been in contact with them, the regulation of sales and markets, and the prohibition of movement of swine from certain districts.[75] At the same time, the Agricultural Department tried to find a lawful way to do research on swine fever inoculation. In 1886, Richard Paget, MP for East Somerset, asked whether progress had been made in the veterinary investigations of the Agricultural Department with regard to inoculation with attenuated virus in cases of swine fever. John Manners, chancellor of the Duchy of Lancaster,

---

72. Roy to Asquith, October 27, 1893, HO 144/315/B7414A.

73. Draft of a letter to Roy, November 20, 1893, HO 144/315/B7414A; Memo to E. J. Stapleton, n.d., HO 144/315/B7414A.

74. Letter on behalf of the Home Secretary to Roy, November 22, 1893, HO 156/8, 253.

75. Board of Agriculture, *Report of the Departmental Committee appointed by the Board of Agriculture to Inquire into Swine Fever: with Minutes of Evidence, Index, and Appendices* (London: H.M. Stationery Office, 1893), 4.

replied that the scientific investigations had not yet commenced, blaming it on the difficulty experienced in obtaining premises suitable to the requirement of the Act and the necessary license for the performance of the experiments. He added that the department was in communication with the Home Office to obtain the necessary permits.[76]

In 1888, the Agricultural Department asked Home Secretary Matthews whether it could employ veterinary surgeons in inoculating with swine fever. In particular, it wondered whether it would be possible to "grant licenses forthwith to these professional men without the trouble and delay which are incurred in filling up and obtaining signatures to the usual forms."[77] The Home Office informed the department that if the object of the inoculation was "the prosecution of inquiry as to the value and methods of protection against swine fever" a license application was necessary, but since the object was to protect the animals "as a means of prevention or remedy" licenses were not required.[78] For the first time in this debate, the objective of the department's operations, the classification of its project, determined whether the inoculation action came under the Act's oversight or not.

The 1894 and 1895 Cases: The Purpose Test

The exact details of the 1894 Case are unknown, but it was concerned with the "novel process of inoculation and bleeding horses to procure the curative and prophylactic serum called diphtheria antitoxin."[79] The Law Officers provided their opinion as to whether the production of diphtheria antitoxin was an experiment under the Act. The attorney general and future chancellor R. T. Reid and Solicitor General Frank Lockwood concluded: "It would be an experiment to adopt this process for the purpose of ascertaining whether it is a scientific truth. It would not be an experiment to adopt it for the purpose of curing individuals in the belief that it is an ascertained or probable cure."[80] This was the first time in which the object of the operation determined for the law officers whether it

---

76. 308 Parl. Deb. (3rd ser.) (1886) 778.

77. Privy Council Office of the Agricultural Department to the Under Secretary of State, June 12, 1888, HO 144/303/B4345.

78. Lushington to the Clerk of the Council in the Agricultural Department, October 25, 1887, HO 144/303/B4345; Memo by Godfrey Lushington, June 14, 1888, HO 144/303/B4345; Letter on behalf of the Home Secretary, June 19, 1888, HO 156/4, 303.

79. Epitome of Previous Opinions, HO 45/11092.

80. R. T. Reid and Frank Lockwood, Diphtheria Antitoxin: Opinion of Law Officers, November 7, 1894, in Minutes of the Physiological Society Annual General Meeting, January 26, 1895, WA, AMS/MF/88.

was an experiment in its legal meaning. In the case of diphtheria antitoxin, the Law Officers contended that the process was not "calculated to give pain" if the operation was done properly. However, they acknowledged that the operation might be carried out "in such a way as to cause pain, e.g. by drawing an excessive quantity of blood from the same horse." The Law Officers therefore determined that if the process involving the injection of toxins was an experiment by the terms of its aim (ascertaining a scientific truth), it would require a license.[81]

Six months later, in June 1895, legal assistant undersecretary Henry Cunynghame asked to "obtain the opinion on the question therein stated as to the performance of inoculations on animals without a license."[82] Attorney General Reid delivered his opinion on the case—precise details of which are missing—a month later. In contrast to the 1894 Case, Reid concluded that in the current case the experiments were calculated to give pain. He repeated the distinction he made in the previous opinion between two sets of inoculation experiments:

> Inoculation for the purpose of testing the efficacy of a substance known to have cure for disease and produced for the purpose of being used in particular cases, is not an experiment within the meaning of the Act. Also that inoculation for purpose of testing the existence of disease in a particular person is not an experiment. We desire however to point out that the circumstance which prevents these inoculations being experiments is the fact that they are performed upon each particular occasion with a view to the cure of individuals. Wherever the ascertainment of scientific knowledge is in any degree an object, the inoculation is an experiment.[83]

Reid stressed in the opinion that the relevant distinction was between the cure of individuals and the ascertainment of knowledge. The testing of antitoxin done for the cure of individuals should not be restricted by the Act.

The 1896 Case: Any Testing Is an Experiment and Pain Should Be Acute

In the summer of 1895, a new government was elected, and the conservative Matthew Ridley was appointed as home secretary. In January 1896, Ridley

---

81. Reid and Lockwood, Diphtheria Antitoxin: Opinion of Law Officers, November 7, 1894.

82. Cunynghame to unknown, June 15, 1895, HO 156/9, 355.

83. R. T. Reid, Case as for Performance of Inoculation on Animals without a License under 39 & 40 Vict. c. 7: Opinion, July 2, 1895, HO 45/11092.

submitted a new inoculation case to the Law Officers Department. Richard Webster and Robert Finlay were the new attorney general and solicitor general, respectively. They opened their opinion with an assertion that experiments that were performed "for the purpose of producing a particular substance, not of gratifying curiosity enlightened or unenlightened," were not experiments according to the Act.

While the previous Law Officers had distinguished operations of testing or producing for individual cases from the attainment of knowledge, Webster and Finlay drew a line between production and testing. They dismissed the former Law Officers' test of the object of experiment as dividing between inoculation operations that were under the Act and those that were not. They ruled that any testing of inoculation whether for individual cases or "mere idle curiosity" was under the Act if it was painful:

> It seems to us impossible to say what is done is not an experiment for the purpose of testing in each case the quality of the serum, or the nature of the disease. It is an experiment for medical purposes in the ordinary sense of the word, and we are unable to find anything in the Act to shew that its operation was to be confined to experiment made either for the advancement of knowledge in general, as distinguished from knowledge in a particular case, or for the purpose of mere idle curiosity.

According to this legal opinion, the testing of antitoxin—rather than producing it—was an experiment in the meaning of the law. The next question was whether those experiments were "calculated to give pain." In contrast to the Greenfield Case and the opinions of 1895, and in line with the ideas promulgated in the 1892 case, Webster and Finlay asserted that inoculation experiments were not painful in the sense of the Act, but were more of a discomfort:

> Whether the operations in question are 'calculated to give pain' is in strictness in each particular case, a question of facts. But we think that a general principle may be laid down which ought to be adhered to in determining this question . . .
>
> The Act seems to have been directed *against causing acute pain* to animals as by the use of the knife, or other surgical instrument, or of corrosive poison. We cannot think that inoculation producing a disease of more or less gradual development attended with *such discomfort as a child may suffer* from measles or typhoid is within the scope of the Act. We must however add that the decision on this point would in each case

necessarily turn upon the form in which the findings as to the facts is put.[84]

In other words, according to Finlay and Webster, the "classical" vivisection operation of cutting or poisoning was painful, whereas the pain produced by disease was negligible. The Act, they reasoned, referred only to acute pain. In the last part of their opinion, Finlay and Webster suggested that since the issue was "no doubt of great public importance," and in light of the "necessity, for the purpose of any decision of value in settling the law, of having the facts fully and fairly brought out," the Solicitor to the Treasury should "take control over any such proceedings."[85]

When the Home Office communicated the new decision to experimenters, the Law Officers' guidelines about the acute degree of pain were omitted, and the responsibility of the experimenter to detect pain was emphasized. In April 1896, legal assistant undersecretary Cunynghame sent a clarification in the name of the home secretary to a list of recipients about the production of antitoxin serum, "which he is informed are now carried out on a considerable scale." Cunynghame explained that serum production was not an experiment, but the testing of its efficacy during the various stages of production was an experiment—thereby making anyone engaged in the production of antitoxin potentially obliged to register their laboratories and apply for licenses. Whether the experiments were painful or not was a "question of fact in each case. It is therefore incumbent on any gentleman who proposes to carry out such testing processes to satisfy himself that such processes would be in the case of his own experiments unaccompanied with pain." If the experimenter believed that the animal would suffer pain, he would have to ask for a license.[86]

Inoculation in Tables

The Home Office administration kept conveying its doubts about the severity of inoculation experiments even when legal opinions stated that they were painful. The annual reports to the Parliament were revealing in this sense. Between 1878 and 1890, the inspectors organized the experiments into two categories,

84. Richard E. Webster and Robert B. Finlay, Opinion, January 29, 1896, HO 45/11092; emphasis added.

85. Richard E. Webster and Robert B Finlay, Opinion, January 29, 1896, HO 45/11092.

86. Cunynghame to the Secretary of the AAMR, April 27, 1896, HO 156/9, 630. A list of additional recipients of this letter is on p. 635.

one including "the total number of experiments performed during the same period and their general nature," and the other "the number and nature of such experiments in which there is reason to believe that any appreciable pain was inflicted." In his report for 1887, inspector Erichsen introduced a separate table in the returns "showing the number and nature of the experiments performed by each licensee during the year 1887." The table (Table III) included the names of licensees, the certificates they held, the nature of their experiments (physiological, pathological, therapeutical) and a column on "pain." Under the latter, the inspector indicated, for example, "10 rabbits" or "3 frogs."[87]

When inspector George Poore prepared his first annual report and returns, he omitted the category of painful experiments. Starting from his report for 1890 and for the next five years, Poore's report analyzed experiments "classified and arranged according to their general nature" and eliminated his predecessor's reference to the existence of pain. Poore also asked to remove the column "Pain" from Table III, but, confronted with Home Secretary Matthews's objection, he modified it to "Remarks," under which he wrote notes such as "painless," "hypodermic injections," and "inoculations."[88]

In Poore's 1896 return, he divided Table III into (A) "experiments other than those of the nature of inoculations, hypodermic injections, or similar proceedings" and (B) "experiments of the nature of inoculations, hypodermic injections, or similar proceedings."[89] In the report, Poore also listed the diseases about which knowledge increased due to inoculation experimentation. Other than demonstrating Poore's dismissive attitude toward animal pain (see more in chapter 4), the changes in the classification of the experiments, and in particular the replacement of "painless" and "painful" categories with those of "inoculation" and "all the rest," reflected the Home Office's view that experiments of the kind of inoculation were substantially different from other physiological operations.

In Poore's annual report for 1897, he commented that inoculation "is inadequately provided for" in the Act. Poore restated a claim often heard, that recording inoculation experiments under Certificate A was misleading: "It would be cruel, rather than otherwise, to anesthetize an animal before subjecting it to the trivial operation of a prick with a needle."[90] But John Swift MacNeill, MP for

---

87. Experiments on Living Animals, Parliamentary Papers, 1888 (186), 18.

88. Royal Commission on Vivisection, *Final Report*, 10.

89. Experiments on Living Animals: Return Showing the Number of Experiments Performed on Living Animals During the year 1896, under Licenses Granted under the Act 39 & 40 Vict. c. 77, distinguishing painless from painful experiments, Parliamentary Papers, 1897 (239).

90. Experiments on Living Animals: Return Showing the Number of Experiments Performed on Living Animals During the year 1897, under Licenses Granted under the Act 39 & 40 Vict. c. 77, distinguishing painless from painful experiments, Parliamentary Papers, 1898 (215), 4.

South Donegal, thought that the misleading part of the annual reports was not the classification of inoculation under Certificate A, but rather the distinction between painless and painful, which he found "fallacious and deceptive." He agreed that inoculation itself was "practically painless"; but called attention to its results, which he compared to the state of an animal recovering from surgery under anesthesia: "We all know the painful after-results of the administering of anaesthetics. During the time of the actual operation, it may be described as 'painless,' but the pain that follows consciousness is fearful and excruciating." MacNeill added that these experiments were "most eminently painful—for living animals are given such diseases as cholera, diphtheria, small-pox, and many other painful diseases. These are described as painless experiments, because they are painless at the time; but I do not think that the public ought to be deceived in this manner."[91]

Attempts to define experiments were still complicating decision-making, and even the identity of the operator was considered a relevant criterion. Inspector Russell shared his perspective with inspector George Thane that "no definition of experiment is given in the Act but it appears to mean some act or proceeding applied to a living animal the result of which is not certainly known to the author of the experiment or to the onlooker." This is why, he reasoned, demonstration of an operation on a living animal might not be considered an experiment from the point of view of the teacher, but it will constitute an experiment "because of the ignorance of the students." Preparation of antitoxin, he continued with this logic, was not considered an experiment since "now the process is well-known," while its testing at different stages was still under the Act.[92]

Notwithstanding the critique and the legal opinions, during a parliamentary debate in 1898, Home Secretary Ridley remarked, "I do not know whether inoculation is really an experiment, but it is so considered."[93] Inspector Thane's report for the same year categorized inoculation experiments as painless, and used the controversial expression when explaining that the pain inflicted by inoculation experiments was "not greater than the prick of a needle."[94] Stephen Coleridge from the Victoria Street Society then published an open letter to the newly appointed Home Secretary Charles Ritchie: "The assurance that no pain can legally be inflicted beyond that of the prick of a needle under this certificate

---

91. 56 Parl. Deb. (4th ser.) (1898) 1512.

92. James Russell to George Thane, August 8, 1899, HO 144/445/B30018.

93. 63 Parl. Deb. (4th ser.) (1898) 544.

94. Experiments on Living Animals: Return Showing the Number of Experiments Performed on Living Animals During the year 1899, under Licenses Granted under the Act 39 & 40 Vict. c. 77, distinguishing painless from painful experiments, Parliamentary Papers, 1900 (211), 4.

are simply false."[95] A Home Office memo signed by W. P. D suggested the following response: "It is perhaps an unhappy remark and might be omitted; people want to know about the consequences of the operation, not the operation itself." Furthermore, in relation to Certificate A, which "states that to use anaesthetics would frustrate the object of the experiment" the memo writer added in brackets: "this statement is nearly always false: but that point is not raised."[96]

Whereas in previous years the Home Office had delivered its instructions to its list of registered licensees and the AAMR, by the mid-1890s new actors from the commercial scene joined the communications. A copy of a letter was sent to inspector Poore, asking him to circulate the new instructions, mentioning specifically a "firm of druggists at Dartford," which was preparing antitoxin serum.[97] The emblem of the new market was Silas Burroughs's and Henry Wellcome's company.[98]

The Wellcome Case: Inoculations Are Painful

In 1900 Burroughs, Wellcome and Co. applied to register Wellcome Physiological Research Laboratories as a place of vivisection. The novel application raised two main concerns at the Home Office. The first was whether a commercial place could be registered under the Act given the policy not to allow the registration of private laboratories. The second concern was whether the operations of Wellcome Laboratories, including the production and standardization of antitoxins, were "experiments calculated to give pain," which necessitated registration. Kenelm Digby, a county court judge who in 1885 succeeded Lushington as a permanent undersecretary, sent a preliminary letter to Finlay, now the attorney general, about this "very important precedent." Digby wished to know whether the Wellcome Laboratories case should be formally submitted to the Law Officers. He was hoping that Finlay, despite his view in the 1896 case that any testing of inoculation was an experiment, would say that the operations undertaken by Wellcome Laboratories did not require licensing and registration. This could have relieved the Home Office from delving into the contentious status of commercial laboratories.[99]

---

95. An Open Letter to the Right Hon. The Secretary of State, December 1900, HO 144/606/B31612.
96. Memo, W.P.D, February 15, 1901, HO 144/606/B31612.
97. Cunynghame to Poore, April 17, 1896, HO 156/9, 626.
98. Cunynghame to Burroughs, Wellcome and Co., May 4, 1896, HO 45/11092.
99. Kenelm Digby to the Attorney General, January 7, 1901, HO 45/11092.

Digby, less legalistic and formalistic than his predecessors at the office, contended that the status of operations "necessary for ascertaining the properties of commercial products" was worth reconsidering.[100] The main question to be answered was whether the Wellcome experiments were painful. He explained that most operations conducted at Wellcome were with diphtheria antitoxin, of which "there is no sufficient evidence of the operation being 'calculated to give pain' to require or even to justify the Home Secretary in taking proceedings."[101]

Drawing from the 1896 opinion that specified "acute" pain as being the Act's threshold ("The Act seems to have been directed *against causing acute pain* to animals as by the use of the knife, or other surgical instrument, or of corrosive poison"), Digby understood the Act's "pain" to be "such pain as in a human requires the use of anaesthetics." Digby clarified that this was not the pain that accompanied inoculation experiments, which entailed "mere discomfort of languor or lassitude or struggles such as usually accompany dying." But while diphtheria's pain was considered unremarkable, tetanus entailed "substantial pain."[102] (That was the reason Wellcome Laboratories had postponed its testing of tetanus antitoxin pending the decision on their registration application.)[103] This led Digby to his concluding question for the Law Officer: was the home secretary allowed to ignore the performance of tests when it was "not in public interest" to enforce the law upon them?[104]

To assist the Law Officers, the Home Office produced a Memorandum Embodying Results of Enquiry made by the secretary of state as to whether Pain is caused by the Inoculation of Animals with Diphtheria Toxin and Antitoxin.[105] The authors of the document are unknown, and there is no indication of similar reports composed beforehand. The memorandum included a description of the process done to ascertain the toxicity of a toxin. Taking a guinea pig as an example, it stated that a brief rise in temperature immediately followed the injection of the toxin, which was succeeded by a steady fall. According to the memorandum, it was due to the body's low temperature and the coldness of the night that a greater number of inoculated animals died at night.

---

100. Digby's attitude is explained by Pellew's notion that his appointment in the Home Office coincided with a more pragmatic and flexible era at the Home Office, "less wedded to precedent and legalistic interpretation." Pellew, "Law and Order," 69.

101. Kenelm Digby to the Attorney General, January 7, 1901, HO 45/11092.

102. Kenelm Digby to the Attorney General, January 7, 1901, HO 45/11092.

103. Case For Opinion, HO 45/11092.

104. Kenelm Digby to the Attorney General, January 7, 1901, HO 45/11092.

105. Memorandum Embodying Results of Enquiry made by the Secretary of State as to whether pain is caused by the Inoculation of Animals with Diphtheria Toxin and Antitoxin, HO 45/11092.

The moment of injection was followed by a swelling, "but at first the animal does not show signs of pain when touched there." Local necrosis (cell injury) sometimes developed in a few days, together with sloughing of the skin, "but the experiment is in most cases completed before this and the animal ought to be killed." In cases of a fatal dose, "the animal at first becomes lethargic it breathes quickly and remains huddled up, it refuses food, which is very unusual for a guinea pig, its hair is puffed out; its eyes are dull; and it loses in weight; it is not restless, as guinea pigs are, when approached; it is obviously very ill; occasionally it shivers."[106] Finally, the inoculated guinea pig lies on its side, opening and shutting its mouth, and dies.

Finlay signed his opinion in February 1901. To the dismay of the Home Office administration, he restated his earlier view that testing and standardizing drugs and antitoxins were experiments. These experiments should not be prohibited, but may be conducted under a license:

> They are performed for the purpose of obtaining knowledge of the quality of the particular specimen of drugs which will determine whether or not it will be useful for saving life, or alleviating suffering. No doubt the words of sec. 3(1) appear to be rather applicable to general scientific discovery than to the ascertaining the quality of a particular specimen of drug, but I think that the words are capable of a construction which would admit of the license; indeed it would reduce the Act to an absurdity if operations so necessary and so beneficial were absolutely prohibited.[107]

As for the question of pain, Finlay no longer thought inoculation produced "discomfort as a child may suffer." After reading the memorandum, he noted that the symptoms in diphtheria antitoxin experiments went "far beyond" those in the case leading to his 1896 opinion. Finlay also added a note about enforcing the Act. He clarified that the home secretary had no duty to prosecute, yet remarked, "it would be very difficult on public grounds to justify entire inaction if he had officially knowledge that the Act was being extremely violated."[108]

The attempts to fit the Act to the emerging bacteriological research sometimes led to absurdities. In the 1890s, disease research made its way into the teaching curriculum, confronting Home Office officials with additional

---

106. Memorandum Embodying Results of Enquiry made by the Secretary of State as to whether pain is caused by the Inoculation of Animals with Diphtheria Toxin and Antitoxin.

107. Finlay, Opinion, February 12, 1900 (should be 1901), HO 45/11092.

108. Finlay, Opinion, February 12, 1900 (should be 1901), HO 45/11092.

difficulties. Subsection 3(5) of the Act stated that the "experiment shall not be performed as an illustration of lecture in medical schools, hospitals, colleges, or elsewhere." The Act allowed illustrative experiments with a special certificate, when anesthetics were in use and when "the proposed experiments are absolutely necessary for the due instruction of the persons to whom such lectures are given with a view to their acquiring physiological knowledge or knowledge which will be useful for them for saving or prolonging life or alleviating suffering." The advisers at the Home Office interpreted the provision as allowing an experiment in illustration of a lecture only when performed on an animal "which is under the influence of some anaesthetics of sufficient power to prevent it from feeling pain, and that the state of insensibility to pain thus induced must be maintained throughout the entire experiments," even if the licensee obtained certificates that allow him to dispense with the use of anesthesia or to allow the animal to recover from anesthesia.[109]

In consideration of these restrictions, inspector James Russell at first tended to decline the request of Robert Muir, a Scottish physician and pathologist, to attain in 1894 a Certificate C for a "simple subcutaneous or intra-venous injection of the organisms of tubercle, Diphtheria, Anthrax, Actinomycosis, and Suppuration, to illustrate the manner in which these diseases are produced," using rabbits and guinea pigs. Russell reasoned that anesthetics "seem necessary in order to comply with the Act though their use is not dictated by common sense."[110] Russell asked inspector Poore for instructions and was provided with Lushington's memoranda on the topic. The memo clarified that the home secretary "has always treated inoculations as experiments and he could hardly make a special exception of inoculations before a class; neither could he reverse the rule: there is very substantial reason why inoculations generally should be treated as pain-giving experiments, viz. that they do as a matter of fact give considerable pain by imparting a serious illness." The Act, Lushington thought, should be amended to accommodate the problem of inoculation better. Meanwhile, the proposed solution was to divide inoculation demonstrations into two separate operations: one of the injection, to be done under anesthesia in the class illustration, killing the animal before recovery. The second step was to inoculate an animal without anesthetics before presenting it to the students and later demonstrate the procedure's after-effects. "This may not seem very reasonable conclusion, but it secures compliance with the law, and the worst result will be that a

---

109. Experiments (Wholly or Partially without Anaesthesia) in Illustration of Lectures, W.P. Byrne to the Secretary of the Royal Commission on Vivisection, June 8, 1909, HO 114/2.

110. Russell to Poore, December 31, 1894, HO 144/959/B7488.

double set of animals will be used, but there will be no serious interferences with instruction."[111] Russell then concluded that Muir could inoculate animals in his laboratory and show them to a class, whereas "should he ever desire to show the use of a hypodermic needle or the operation of scratching the skin he can apply for Certificate C and kill the animal under anaesthesia."[112]

The disputes surrounding what kind of pain should be recognized by law had a central part in the Act's formation, both by charting which scientific practices were under state supervision and shaping the relationship between the actors involved. During the debates over inoculation, scientists and lawyers remade their territory of expertise. Each was implicated in defining the terms of the others' practices. While the lawyers decided upon issues at the core of the scientific work, the experimenters argued that the lawyers' interpretation of the Act was misguided and harmful to animals. During these debates, legal and scientific actors reversed roles as physiologists provided interpretations of the Act and its aims, while Law Officers set definitions for experiments and pain.

Although the Home Office's mandate was to supervise and manage animals' pain in laboratories, this could not have been done without reexamining and deciding what counted as an experiment. Defining pain and experiments had a direct impact on the reach of the law and on the Home Office's workload. It affected the statistics of animal experimentation published by the office: the broader the definitions, the higher the number of experiments reported, and the higher the numbers, the greater the public critique. From the perspective of the scientists, at stake was their reputation as well as the legality of their deeds. While bureaucrats defined a core concept in scientific practice, experimenters provided interpretation for the law, drawing from their claimed privileged access to animal pain. From the perspective of the civil servants at the Home Office, they had to navigate between a promising new science and an Act that was almost irrelevant concerning disease research and antitoxin production.

Inoculation experiments, such as Yeo and Ferrier's brain research discussed in chapter 2, challenged the claim that science had the best tools to solve the problems it created. Once scientists applied to the Home Office to experiment without anesthesia, the most powerful argument they had so far in favor of animal experimentation turned against them. Lawyers and state bureaucrats then attempted to make regulation—rather than medicine—the means by which

---

111. G.L., Memoranda re Certificate C with Regard to Inoculation, December 14, 1894, HO 144/959/B7488.

112. Russell to Poore, March 2, 1895, HO 144/959/B7488.

pain is reduced to a minimum. The inoculation cases discussed above also reveal the significant role of lawyers in defining pain and experiment. Four pairs of Law Officers produced six opinions, most of which contended that inoculation experiments were experiments calculated to give pain. At the same time, the Home Office's resubmission of the cases for the opinion of the Law Officers, even after an opinion on a similar case was provided, shows that the repeated objections of experimenters to the rationale of the Law Officers convinced the different Home Secretaries of the need for reevaluation.

By the turn of the twentieth century, the pain involved in inoculation was no longer disputed. When the second Royal Commission on Vivisection began its work in 1906, inoculation experiments were the majority of experiments. The commission concluded that "even if the initial procedure in cases under that certificate may be regarded as trivial, the subsequent results of this procedure must in some cases, at any rate, be productive of great pain and much suffering."[113] The commission also agreed with previous legal opinion when distinguishing between experimentation and production, recommending that vaccine and serum production would not be overseen by the Act.[114]

The Home Office's memos and Law Office's opinions regarding bacteriological research show that the threshold of pain, which guided the inclusion of certain experiments in the Act, shifted constantly. However, the setting of "acute" as the standard by the 1896 opinion, and the adoption of this measure by permanent undersecretary Digby was inconsistent with a testimony provided by inspector Russell about his work during the 1890s and early twentieth century. Russell testified that as a policy, the Home Office had required a license to any experiment calculated to give pain, however slight: "the word pain had been taken not in the popular sense, but in a much wider sense, whatever is supposed to cause the slightest discomfort or uneasiness has been taken to be calculated to give pain."[115] The next chapter delves into the inspectors' work, tracing the way they fulfilled the administrative tasks to record and monitor animal pain in research laboratories.

---

113. Royal Commission on Vivisection, *Final Report*, 52.

114. In 1910, about 95 percent of reported experiments were "of the nature of simple inoculation, hypodermic injections and similar proceedings performed without anaesthetics." Royal Commission on Vivisection, *Final Report*, 51.

115. Royal Commission on Vivisection, *Appendix to First Report*, 22.

4

REGULATING PAIN IN LABORATORIES / The Inspectorate

On March 14, 1894, George Vivian Poore (1843–1904) strolled around the Bacteriological Laboratory at King's College. He noted in his book: "Saw a rabbit which had been lately inoculated, with the bacillus of tetanus. The animal presented no abnormal symptoms, hopped about as usual."[1] No other remarkable sights were recorded on that day. In his next visit two months later, Poore, a prestigious physician and an inspector under the Vivisection Act, documented a lethargic rabbit. He also planned to inspect the Neuropathological Laboratory, but it was locked. On the next day, he journeyed to another registered laboratory, where he made sure all required licenses and certificates were duly held.

At the core of the implementation of the Act were the day-to-day activities of the Home Office inspectors and their encounters with animals subjected to research. Section 10 of the Vivisection Act set the legal foundation for the employment of inspectors: "The Secretary of State shall cause all registered places to be from time to time visited by inspectors for the purpose of securing a compliance with the provisions of this Act." The Act also allowed the home secretary to "appoint any special inspectors, or may from time to time assign the duties of any such inspectors to such officers in the employment of the government, who may be willing to accept the same, as he may think fit, either permanently or temporarily."

The inspectors were legal agents with scientific credentials who moved back and forth between the Home Office and the laboratories, navigating between the law's requirements and scientific demands. Moving between the legal and the scientific spheres, they shaped Home Office policies toward animal experimentation and were central to the incorporation of legal norms in the physiological laboratory.

---

1. Dr. Poore's Report on Visitation of Registered Places during 1894, 1895, HO 144/370/B17451A, 4 (hereinafter: Poore's Report on 1894).

Constructing pain as an object of regulation meant that inspectors were constantly engaged in empathizing with animals by delineating parameters for the identification of pain, such as appetite and fever, or the dismissal of pain by referring to the animals' social behavior. The testimonies of Poore and his colleagues about the animals they saw show that they were reluctant to report animal suffering in laboratories even though they closely observed numerous diseased and injured animals. The alleged blindness to animal suffering was due to, among other reasons, the difficulties of defining pain. Instead, inspectors focused on procedural breaches of the Vivisection Act and were proactive in educating the experimenters about the Act's requirements.

Science studies scholars recognize qualified witnessing as fundamental to modern empirical sciences. Through their laboratory visits and remote oversight over animal experimentation, the vivisection inspectors were witnesses in two meanings. They acted as public representatives examining the welfare of animals, and they were also scientific witnesses, instrumental in the coproduction of cruelty-free facts. By directing the physiologists' compliance with the Act, the inspectors intertwined practical and normative aspects of animal research. These processes reshaped the way that animals were housed, interpreted, and treated.

Meet the Inspectors

The first inspector under the Vivisection Act was George Busk (1807–86), a respected surgeon and a naturalist in his seventies. He was a fellow of the Royal College of Surgeons and held the chair of Hunterian Professor of Anatomy and Physiology before serving various offices at the college. Busk was involved with several scientific societies; he was a fellow, and also served on the council of the Royal Society and was a member of the Linnaean, the Geological, and the Zoological societies.[2] The antivivisection journal, the *Zoophilist*, however, contended that Busk had been "remarkable for nothing but mediocrity," and his inspection was "a sham."[3] For tasks that required traveling outside of London, Busk employed an assistant, G. J. Allman.

The surgeon John Eric Erichsen replaced Busk on April 1, 1886. His resume was as distinguished as Busk's. He was a fellow of the Royal College of Surgeons and occupied high positions at the institution. Erichsen was elected a fellow of the Royal Society in 1876, and, in 1877, he was made surgeon-extraordinary

---

2. "George Busk, F.R.S.," *British Medical Journal* 2, no. 1337 (1886): 346.
3. "Exit Busk, Enter Erichsen," *Animal's Defender and Zoophilist* 6, no. 1 (May 1, 1886): 4–5, 4.

to the Queen. Erichsen was a commissioner of the first Royal Commission on Vivisection and was considered a neutral nominee, although he had previously experimented on animals. He even managed to gain the trust of Commissioner Richard Holt Hutton, a leading antivivisection spokesman.[4] In 1881, Erichsen was the president of the surgical section at the meeting of the International Medical Congress in London.[5]

Erichsen did not identify himself with the antivivisection cause. In the pages of the *British Medical Journal* he refuted an antivivisectionist who had tried to use one of his past statements to make a claim against vivisection.[6] Moreover, he did not hide his support for medical research when upon his resignation from the position of council member of the Royal College of Surgeons, he wished that the college's council would "shortly be largely increased by the establishment of physiological and pathological laboratories on the largest and most approved scale in connection with the College."[7] In light of his service on the Royal Commission, and his past experience with vivisection, the *Zoophilist* published a set of denunciating comments upon his appointment.[8] Supporters of vivisection also targeted Erichsen. During a House of Commons debate one MP voiced complaints that Erichsen was "a great deal too strict, that he holds the reins too tightly, and that he will not allow latitude enough."[9]

Inspector Erichsen proposed to assign an additional inspector in February 1888, noting that since the "duties would necessarily be of a somewhat delicate character, requiring much tact as well as judgment for their due performance," it required a careful selection.[10] A correspondence took place between the Treasury, Erichsen, and the Home Office as to whether any fee could be charged for the registration of places under the Act in order to enable the hiring

---

4. French, *Antivivisection and Medical Science*, 96.

5. D'A. Power, "Erichsen, Sir John Eric, baronet (1818–1896)," rev. B. A. Bryan, *Oxford Dictionary of National Biography* (Oxford University Press, 2004), http://www.oxforddnb.com/view/article/8835.

6. John Eric Erichsen, "The Antivivisectionists and the Progress of Modern Surgery," *British Medical Journal* 2, no. 985 (November 15, 1879): 794.

7. Sir John Eric Erichsen, *The Member, the Fellow and the Franchise [in the Royal College of Surgeons]* (London: HK Lewis, 1886), 16.

8. "Exit Busk, Enter Erichsen," *Animal's Defender and Zoophilist* 6, no. 1 (May 1, 1886): 4; "Mr Erichsen, The New Inspector, as a Vivisector," *Animal's Defender and Zoophilist* 6, no. 2 (June 1, 1886): 36–37; "The Anti-Vivisection Attack on Mr. Erichsen," *British Medical Journal* 1, no. 1482 (May 25, 1889): 1180.

9. 336 Parl. Deb. (3rd ser.) (1889) 614.

10. Erichsen to Home Secretary Henry Matthews, February 6, 1888, HO 144/299/B2697.

of an additional inspector.[11] The Law Officers of the Crown rejected the idea of charging fees, maintaining that the Act gave them no power to do so.[12]

Exhausted from traveling to visit the laboratories "scattered over the whole length of England and Scotland, from Plymouth to Aberdeen and from Bristol to New Castle-on-Tyne," Erichsen presented an ultimatum: with no assistance, he would resign.[13] Home Secretary Henry Matthews, who was "very anxious not to lose the valuable service" of Erichsen, pressured the Treasury for additional support.[14] Finally, the Treasury added an annual sum of £250 for the salary and traveling expenses of an additional inspector.[15]

Poore first joined the Vivisection Act's administration on April 1, 1889 to assist Erichsen.[16] The latter strongly supported Poore's nomination, who among his other advantages was "well known in the medical profession and has never been engaged in caring experiments on living animals."[17] Poore had the right profile for the job. A member of the medical elite, he had been a fellow of the Royal College of Physicians since 1877. He also had a special interest in public hygiene and was a member of the Executive Council of the International Health Exhibition held in London in 1884. He held the position of physician at University College Hospital, where he was also the chair of medical jurisprudence and clinical medicine.[18]

Newly appointed as assistant inspector, Poore asked for the following supplies: some official paper and envelopes (which would serve him for frequent correspondence with the Home Office and physicians involved with vivisection),

---

11. Home Office negotiations with the Treasury for inspectorate wages were also examined in Jill H. Pellew, "The Home Office and the Explosives Act of 1875," *Victorian Studies* 18, no. 2 (December 1, 1974): 188.

12. Letter on behalf of the Secretary of State to the Secretary to the Treasury, July 27, 1888, HO 156/4, 339–40; Cruelty to Animals Act 1876, Opinion by Edward Clarke and Richard Webster, July 21, 1888, HO 144/299/B2697.

13. Erichsen to Lushington, January 14, 1889, HO 144/299/B2697.

14. Letter on behalf of the Secretary of State to the Secretary to the Treasury, January 25, 1889, HO 156/4, 485.

15. Memo, As to Inspection under 39 & 40 Vict. c. 77., HO 144/960/B8417. The pressure to employ additional inspectors was present in the following years. For example, in 1896, the Victoria Street Society asked the Home Office to increase the number of inspectors. Secretary of State Matthew Ridley declined the request, explaining that "he has no reason to believe that the provisions of the Act . . . are not duly enforced." Secretary of State to the Secretary of the Victoria Street Society for the Protection of Animals from Vivisection, May 4, 1896, HO 156/10, 17.

16. Godfrey Lushington to the Secretary of the Treasury, April 4, 1889, HO156/4, 574.

17. Erichsen to [?], March 26, 1889, HO 144/299/B2697.

18. "George Vivian Poore, M.D., F.R.C.P., Emeritus Professor Of Medicine, University College, London," *British Medical Journal* 2, no. 2292 (December 3, 1904): 1544–46,1544; "The Late Dr. Poore," *British Medical Journal* 1, no. 2321 (June 24, 1905): 1398.

the necessary forms for expenses (mostly related to his travels), and a copy of the Act and any conditions imposed by the home secretary.[19] Soon, he needed more of these bureaucratic building blocks. On May 12, 1890, upon the resignation of Erichsen, Poore was promoted to the rank of inspector.[20] Erichsen provided him with the following instructions: to visit any registered place at least twice a year; that those places where much experimentation was carried out would have to be inspected even more frequently; to be occasionally present at experiments, especially at those that might be involving pain; and to inspect all places proposed for registration before they were licensed.[21]

The workload of the inspectors had steadily increased. In 1888, there were seventy-five licensees and twenty-eight institutions, in which one or more laboratories were licensed to conduct experiments. In 1889, there were eighty-seven licensees and thirty institutions, and, in 1890, the year in which Poore became an inspector, the number of licensees had increased to 110 and the number of institutions to thirty-three.[22] The growth in requests for licenses and registration led Poore to suggest to Home Secretary Matthews that an assistant inspector be engaged with the responsibility of Scotland and Newcastle, leaving England south of York in his mandate as the principal inspector.[23] Poore's request was granted, and James Alexander Russell (1846–1918) was hired as assistant inspector for Scotland and North England starting August 1, 1890.[24]

Russell was trained in medicine at the University of Edinburgh, where he was also an assistant and demonstrator of anatomy. Two years of studying sanitation in France qualified him to engage in public health management. In 1880, he joined the town council of Edinburgh, and later became the chair of its health committee. In 1881, he was elected a fellow of the Royal College of Physicians and, at about the same time, was appointed government inspector of anatomy for Scotland. Russell was engaged, in 1885, in the organization of the Fever Hospital for infectious disease. He was a member of the Royal Society Edinburgh and acted, in the years 1891–94, as the Lord Provost of Edinburgh. Russell was knighted by Queen Victoria in 1894.

---

19. Poore to Godfrey Lushington, April 24, 1889, HO 144/299/B2697A.

20. Inspectors were chosen by nomination and not by competition. Letter on behalf of the Secretary of State to Mr. Jones, December 29, 1891, HO 156/6, 580.

21. Minutes as to the Appointment of an Additional Inspector, May 22, 1890, HO 144/960/B8417.

22. Experiments on Living Animals, 1889, H.C. 114, 3; Experiments on Living Animals, 1890, H.C. 150, 3; Experiments on Living Animals, 1890, H.C. 266, 435.

23. Letter on behalf of the Secretary of State to the Secretary, H.M. Treasury, December 12, 1890, HO 156/6, 47. Records show that an additional assistant was employed to assist George Poore.

24. H. C., "Sir James Alexander Russell, M.D., LL.D., F.R.C.P.E., F.R.S.E., One-Time Lord Provost of Edinburgh," *British Medical Journal* 1, no. 2979 (1918): 163–64.

As license applications and related inquiries piled up at the Home Office, Home Secretary Matthews urged the Treasury in a lengthy plea to increase the inspectors' compensation. Matthews explained that even with the assignment of Russell as an assistant inspector, "the work under present conditions far exceeds what it fell to the duty of the Inspector to perform during the first ten years after the Act came into operation."[25] In consequence, Poore's annual salary for 1891 was raised to £315.[26] A few years into his position, in February 1897, Russell was asked by the Home Secretary to give "extra assistance" to Poore, adding about £50 to his yearly £157.10 salary.[27] The next month, Russell undertook the inspection of all registered places at a distance of at least 100 miles from London, including Liverpool and Manchester.[28]

George Dancer Thane (1850–1930), an anatomy professor at University College in London, replaced Poore on June 28, 1899. Thane was one of the founders of the Anatomical Society of Great Britain and Ireland, an active member of the German and French anatomical societies, and a member of the Royal College of Surgeons. He was a professor of anatomy at University College London and stated to have never experimented on living animals.[29] Thane began visiting laboratories while Poore was still on duty, and the laboratory visits he conducted during his first year in office were executed under the supervision of Poore.[30] When Poore retired, Thane worked next to Russell. Thane was knighted in 1919 for his service as inspector and for his services to University College.[31]

---

25. Letter on behalf of the Secretary of State to The Secretary, H.M. Treasury, December 12, 1890, HO 156/6, 47.

26. Memo, As to Inspection under 39 & 40 Vict. c. 77.

27. Secretary of State to Russell, February 18, 1897, HO 156/10, 505; A letter draft on behalf of the Secretary of State to the Secretary to the Treasury, January 1897, HO 144/960/B8417.

28. Memo, As to Inspection under 39 & 40 Vict. c. 77.

29. Royal Commission on Vivisection, "Appendix to First Report of the Commissioners : Minutes of Evidence, October to December, 1906" (London : printed for H.M.S.O. by Wyman & Sons, 1907), 15.

30. In addition, Poore was still managing Russell. For example, in May 1899, Poore sent Russell a telegram regarding a report by T. F. Fraser. James Russell, *A List of Visits to Registered Places under the Act during 1899*, HO 144/451/B30824, 32 (hereinafter: Russell's Report on 1899).

31. H. A. H., "Sir George Dancer Thane, Ll.D., Sc.D., F.R.C.S., Emeritus Professor of Anatomy, University College, London," *British Medical Journal* 1, no. 3603 (January 25, 1930): 175; "Obituary: Sir George Dancer Thane," *Nature* 125, no. 3147 (1930): 281. This chapter does not include the distinct inspection, much smaller in scope, which was operating in Ireland. The Irish inspector during the period this chapter surveys was William Thornley Stoker. Anatomist and surgeon, the Irish inspector had furnished his younger brother Bram Stoker, author of the Gothic novel *Dracula*, with ideas about physiological manipulations of the brain. Anne Stiles, *Popular Fiction and Brain Science in the Late Nineteenth Century* (Cambridge: Cambridge University Press, 2011), 61.

The inspectors' tasks were to mediate between experimenters and policy-makers, examine all registered places, and report deviations from the law while maintaining congenial relationships with the experimenters to enjoy relatively free access to animal rooms, demonstration rooms, and even the performance of experiments. Examining the work of inspectors affords a sense of the everyday life of the law: It shows how the implementation of the Act was influenced by the persona of the inspectors and their immersion in the scientific world, by the way they approached and scrutinized the animals for signs of pain and by the relationships they cultivated with the experimenters.[32]

"High Influential Position ... and in the Vigour of Life": Electing Inspectors

Because the attempts to regulate vivisection enraged many experimenters and generated major institutional resistance, officials emphasized the importance of ensuring that the inspectors be personally agreeable. The Royal Commission on Vivisection indicated that the inspectors should be "persons of such character and position as to command the confidence of the public no less than that of men of science."[33] During the preparation of the vivisection bill, members of the medical establishment joined forces to mitigate the potential interference in their work, using their ties with Parliament members and the public platform provided by medical journals.[34] The General Medical Council wished to amend the vivisection bill to ensure the employment of inspectors who were "scientifically competent to appreciate the nature and intention of such experiment as he may witness."[35] This recommendation did not make its way into the Act but was unofficially followed by the Home Office administration in practice in order to achieve legitimacy among experimenters.

Home secretaries recognized that the character of the inspectors was crucial in legitimizing their reports in the eyes of experimenters. The Home Office

---

32. I follow the work of scholars who analyzed the relations between British people, veterinarians, and policymakers in response to state monitoring of the use of animals. In particular, Abigail Woods, *A Manufactured Plague: The History of Foot-and-Mouth Disease in Britain* (London: Earthscan, 2004); Neil Pemberton and Michael Worboys, *Mad Dogs and Englishmen: Rabies in Britain 1830–2000* (Basingstoke: Palgrave, 2007).

33. Royal Commission Report, 221.

34. French, *Antivivisection and Medical Science*, 118–42.

35. General Medical Council, "Memorial from the General Medical Council to Her Majesty's Government Respecting the Bill Intituled [*sic*] 'An Act to Prevent Cruel Experiments on Animals,'" n.d., file 2021/7, 5, Royal College of Physicians, London.

sought to establish institutional legitimacy by recruiting inspectors from within the supervised professions, expecting that those with such professional backgrounds could produce credible testimony about the state of animals in laboratories. Home Secretary Matthews maintained in 1890 that the inspectors' duties "cannot be performed except by a gentleman who has high influential position in his profession and is in the vigour of life."[36] In the Vivisection Act's first few decades, the Home Office recruited inspectors with medical training, who were also involved in research and served in educational or public institutions related to their scientific expertise.[37] The scientific background of the inspectors played a role in making them agents of coproduction; being scientists who communicated with other scientists, they facilitated the flow of information about animals' bodies between the Home Office and physiologists. This flow also carried with it norms of conduct.

Inspectors were chosen by nomination rather than by competition, and until 1912 were employed on a part-time basis.[38] The stated aim was to find gentlemen who would be viewed as credible by both the scientific community and the wider public, which was envisioned to include many critical perspectives on vivisection. However, it is evident that the satisfaction of the scientific community was a top priority. Home Secretary Herbert Asquith defended the importance of scientific training for the inspectors and said that while the Act only required the inspection of a registered place, "the inspector would not be doing his duty properly were he not to make himself personally acquainted with the matter in which the experiments themselves were conducted." Asquith added that inspectors "ought not however to look on the eminent scientific men who conducted these experiments as if they were criminals caught out at attempted infractions of the law."[39] It is hard to overstate how meaningful this last notion was for the scientific community. The original skepticism of physiologists gradually decreased when it became clear that inspectors would have ties to the scientific enterprise. Therefore, the experimenters' discontent with the administration of the Act was aimed mainly at the home secretary rather than at the inspectors. The pathologist Charles S. Roy from the University of Cambridge remarked

---

36. Secretary of State to the Secretary, December 12, 1890, 50.

37. Veterinary doctors, who had a lower status in the scientific milieu at that period, were qualified to serve as inspectors much later in the twentieth century (Home Office 1980, 12).

38. Letter on behalf of the Secretary of State to Mr. Jones, December 29, 1891, HO 156/6, 580. The Home Office was reluctant to accept open competitive entry for recruits, arguing that civil service positions were trust based. Pellew, *The Home Office*, 20–22. See also Sydney Littlewood et al., "Report of the Departmental Committee on Experiments on Animals" (London: Her Majesty's Stationery Office, 1965), 41.

39. "Memo on Inspection under the Cruelty to Animals Act," 1902, HO 144/634/B37080.

that the first three inspectors "have admirably fulfilled their duties as inspectors." Nevertheless, he argued that their mandate in advising the home secretary should be limited, as "none of them had the laboratory experience required to enable them to act as efficient experts."[40]

The requirement that inspectors have a scientific background also reflected the emerging centrality of scientific knowledge in defining British attitudes toward animals. Sir William Church, the president of the Royal College of Physicians, voiced the position held by scientific institutions when he wrote in 1899: "I feel certain that if a layman were appointed inspector there would be a general uprising of the profession of scientific teachers and men throughout the kingdom." Church explained that by layman he meant "a lawyer or other person who had not gone through a complete scientific training including either physiology or medicine."[41] Church specified that the inspectors should not only be men of science but also trained in specific fields. In his view, a background in physiology and medicine was necessary to accurately understand the tasks demanded of experimenters in laboratories. Mastery of these disciplines was also required to produce a credible evaluation of the suffering of animals in laboratories. The inspectors' medical training qualified them as expert witnesses of animal pain.

The Vivisection Act's opponents maintained that the inspectors were too closely attached to their subjects of supervision and too embedded with the values, interests, and goals of the scientific establishment to ensure a disinterested inspection. In a debate at the House of Commons in September 1893, Colonel Amelius Lockwood, a conservative MP for Epping, claimed that the inspectors "were supposed to stand between the vivisectors and their subjects, the dumb animals, in order to assure . . . the claims of the lower animals to human consideration." MP Lockwood argued that the inspectors were insufficient in this regard—first, because they held other occupations in addition to being inspectors and, second, because they were not sufficiently detached from the subject. Lockwood pointed out that Inspector Poore, whom he acknowledged to be "universally respected," had held a position at the University College Hospital and that "a gentleman who was constantly in association with 16 men engaged in the work of vivisection could hardly be regarded as perfectly impartial." Lockwood recommended that future inspectors be appointed in consultation with the Royal Society for the Prevention of Cruelty to Animals. Home Secretary Asquith smiled when he heard Lockwood's proposal: "I am afraid that if I had to arrive at

40. Charles S. Roy to the Secretary of State, October 28, 1893, HO 144/315/B7414A.

41. "Memo on Inspection."

a conclusion with the Anti-Vivisection Societies we should remain without any inspector at all."[42]

Asquith's successor at the Home Office, Matthew White Ridley, had to confront a similar critique on the scientific background of the inspectors. In an 1897 debate in the House of Commons, MP MacNeill asked bluntly whether the inspectors were advocates of vivisection. Ridley replied that the inspectors had been appointed without regard to their advocacy of, or opposition to, vivisection. He argued that "the scientific knowledge required in an inspector implies a knowledge of experiments on animals," and added that he was persuaded that "you cannot possibly examine and test whether operations are conducted according to law unless you employ professional gentlemen who know something about the matter."[43]

The strategy of employing familiar and respected figures in the scientific community who were versed in the medical sciences evoked undesired, perhaps even unexpected, opposition. The historian of science Steven Shapin describes how, for a long period, scientists had enjoyed a privileged position in relation to public affairs that was later transformed into "moral equivalence."[44] However, the historians Anita Guerrini, Coral Lansbury, and Harriet Ritvo reveal that many British individuals distrusted medical institutions for their abusive treatment of lower-class women and their use of poor populations.[45] For late nineteenth-century criticizers of vivisection, being a man of science not only was unrelated to having a special moral standing but also was seen as disqualifying a person from expressing views on public issues.

The Vivisection Act's administration demonstrated the classical problem of regulatory capture, whereby the regulator agents have an interest in protecting the very entity they are charged with regulating. The historian Jill Pellew notes that, perhaps more than any other inspectorate at that time, the vivisection inspectors were "closer to those they were inspecting than to their official colleagues" in the Home Office.[46] An inspector in such a position, Pellew suggests, might have "overlooked some small infringements of the Act in the interests of his own profession."[47] At the same time, what critics of the Act's administration might have interpreted as an unduly intimate relationship between inspectors

42. 17 Parl. Deb. (4th ser.) (1893) cols. 333, 334, 344.

43. 47 Parl. Deb. (4th ser.) (1897) 1181.

44. Steven Shapin, *The Scientific Life: A Moral History of a Late Modern Vocation* (Chicago: University of Chicago Press, 2010), 23–46.

45. Guerrini, *Experimenting with Humans and Animals*, 87–92; Lansbury, *The Old Brown Dog*, x; Ritvo, *Animal Estate*, 163.

46. Pellew, *The Home Office*, 198

47. Pellew points to anatomy inspectors and vivisection inspectors as the most prone to side with the institutions they were inspecting. Pellew, *The Home Office*, 198.

and experimenters might also have been a way to ease potential tensions as they fulfilled their mission.

There were inherent complexities in the inspectors' role because they interacted with people whose work they supervised and whose careers they could damage. Although the licensees were obliged by law to allow the inspectors to examine the registered places where they operated, a cooperative attitude benefited the inspectors' work. As in other inspectorates, a substantial part of the inspectors' role in relation to the experimenters was explanatory and advisory.[48] The historian Gerald Rhodes observes in his study of inspectors in late nineteenth-century factories and mines that "[they] relied on persuasion rather than prosecution."[49] Along the same lines, the historian Tom Crook claims in his study of sanitation inspectors that "public diplomacy was crucial" and relations with the public determined whether the inspectors' jobs would be relatively easy or "slow and confrontational."[50] When Poore, for example, visited the Pathological Department at Cambridge University, James Lorrain Smith asked him if putting a mouse in a jar in front of a class in order to collect the gases produced through respiration was an experiment. Poore replied negatively, explaining that he did not think it was an experiment calculated to inflict pain.[51] Inspector Thane explained his role to the second Royal Commission on Vivisection as one of familiarizing experimenters with the law: "I go around the laboratory, generally accompanied by a licensee, and I discuss with the licensees what kind of experiments they are doing. I explain to them what the requirements of the Act and of the Secretary of State are (much time is often occupied in giving information, explanation and advice to licensees, or persons wishing to obtain licenses, both personally and by correspondence)."[52]

The Inspectorate's Routine

Inspection was developed as a tool of the British central government in the midnineteenth century. By the time the Vivisection Act was released, there were

---

48. Pellew makes a similar claim in reference to factory inspectors. Pellew, *The Home Office*, 125.

49. Gerald Rhodes, *Inspectorates in British Government: Law Enforcement and Standards of Efficiency* (London: Allen & Unwin for the Royal Institute of Public Administration, 1981), 64.

50. Tom Crook, "Sanitary Inspection and the Public Sphere in Late Victorian and Edwardian Britain: A Case Study in Liberal Governance," *Social History* 32, no. 4 (November 1, 2007): 369–93, 372.

51. Poore's Report on 1894, 1.

52. G. D. Thane, The Royal Commission on Vivisection: Précis of Evidence, HO 144/4.

ten inspectorates working under the Home Office, monitoring a range of areas from factories and mines to burial grounds and salmon fisheries.[53] The concept of inspection was familiar to nineteenth-century men of medicine from the Anatomy Act 1832, and most of the witnesses who testified in front of the Royal Commission on Vivisection expressed no objection to the idea of inspection.

William Fergusson, a sergeant surgeon to the Queen, a surgeon at King's College and a fellow of the Royal Society envisioned the inspection of vivisection as follows: "Just as the common law enables the authorities to send detectives in certain directions to ascertain who may be in certain houses, so they should be able to send men who might say: 'I should like to see the number of dogs and rabbits or cats about this place.' I think that would have a very wholesome effect."[54] The physiologists' responses reflected their experience with the Anatomy Act's inspectors, whose main responsibility was to ensure that dissected human corpses were obtained lawfully. Nonetheless, for the Vivisection Act, the commission rejected the vision of a minimal and formalistic inspection system in the spirit of the Anatomy Act and, in its 1876 report, recommended that greater responsibility be given to the inspectors of vivisection. The commission advised that places where animal experimentation was undertaken should be registered and that the home secretary be given the means to make a "most efficient inspection" of such premises.[55]

The Vivisection Act did not explicitly prescribe any duty other than visiting registered places, but, in practice, inspectors were involved in much more. In addition to laboratory visitations, the inspectors' responsibilities included advising the home secretary, communicating with experimenters, and gathering and organizing data for annual reports presented to Parliament. The inspectors composed reports with direct implications for laboratories. For example, following an 1888 memorandum by inspector Erichsen, the Home Office added a condition to licenses requiring the use of antiseptics in certain procedures, aiming to ease the pain of animals recovering from surgeries.[56] However, the inspectors were not given any authority to begin legal proceedings for violations of the Act,

---

53. Pellew 1982, 12. Gerald Rhodes distinguishes between "enforcement inspectors," who ensure compliance with statutory requirements, and "efficiency inspectors," who supervise certain activities of other authorities. Rhodes classifies the vivisection inspectors as enforcement inspectors. Rhodes, *Inspectorates in British Government*, 3, 247.

54. Pellew, *The Home Office*, 54.

55. Royal Commission Report, 54.

56. Royal Commission on Vivisection, "Final Report of the Royal Commission on Vivisection" (London: H.M.S.O., 1912), 7.

perhaps because of the inflammatory potential of prosecuting physiologists.[57] The home secretary, who held the authority to prosecute, relied heavily on the work of the inspectors in gathering and processing information about the state of affairs in laboratories and designing the office's policies.

The inspectors' visits to laboratories lasted anywhere between two hours and an entire day. For Russell, it was mostly a matter of personal convenience: he was occupied with his additional roles and conducted his visits when it suited his schedule.[58] It was left to the discretion of the inspectors to decide whether to inform the licensees about the anticipated inspections: in 1894, for example, the number of scheduled visits equaled unscheduled visits.[59] In 1895, Russell made all of his visits without an appointment, with one or two exceptions. He explained to Poore that he preferred not to be tied to a day and added that it would improve credibility with antivivisection protestors to be able to report that most inspections took place without advance notice.[60] Visiting laboratories was a delicate task as inspectors needed to move smoothly between the bureaucratic and scientific spheres.

During 1894, for example, Poore paid approximately fifty visits to laboratories in universities and hospitals around England. He traveled to Cambridge, Liverpool, Manchester, Oxford, Netley, Wadsley, Wakefield, and Woolwich. Most of Poore's visits were to laboratories in London, the British center for physiological research: London University College, King's College, St Bartholomew's Hospital, Guy's Hospital, St Thomas' Hospital, St Mary's Hospital, Charing Cross Hospital Medical School, the British Institute of Preventive Medicine, the Royal Veterinary College, the Brown Institution, and the Royal Colleges of Physicians and Surgeons. The next year Woolwich was removed from the list, and a private laboratory on Grove End Road was added.[61] Russell visited laboratories in Edinburgh, Glasgow, Aberdeen, Dundee, and Newcastle: fifty recorded visits during 1894 and fifty-nine during 1895. In 1897, Poore paid sixty-one visits to registered places, and Russell sixty-eight.[62]

The number of visits to each registered place varied. Although Home Secretary Matthews was quoted to have said on one occasion that "the words

---

57. Legislators were well aware of the events surrounding the 1874 case known as the "Norwich Trial," in which an animal protection organization prosecuted physiologists for violating an anti-cruelty law prior to the Vivisection Act. Susan Hamilton, *Animal Welfare & Anti-Vivisection 1870–1910: Frances Power Cobbe* (London: Routledge, 2004), XVI.

58. Appendix to First Report, 22.

59. "Memo on Inspection."

60. James Russell to George Poore, March 5, 1896, HO 144/383/B19846A.

61. Poore's Report on 1894.

62. 62 Parl. Deb. (4th ser.) (1898) 836.

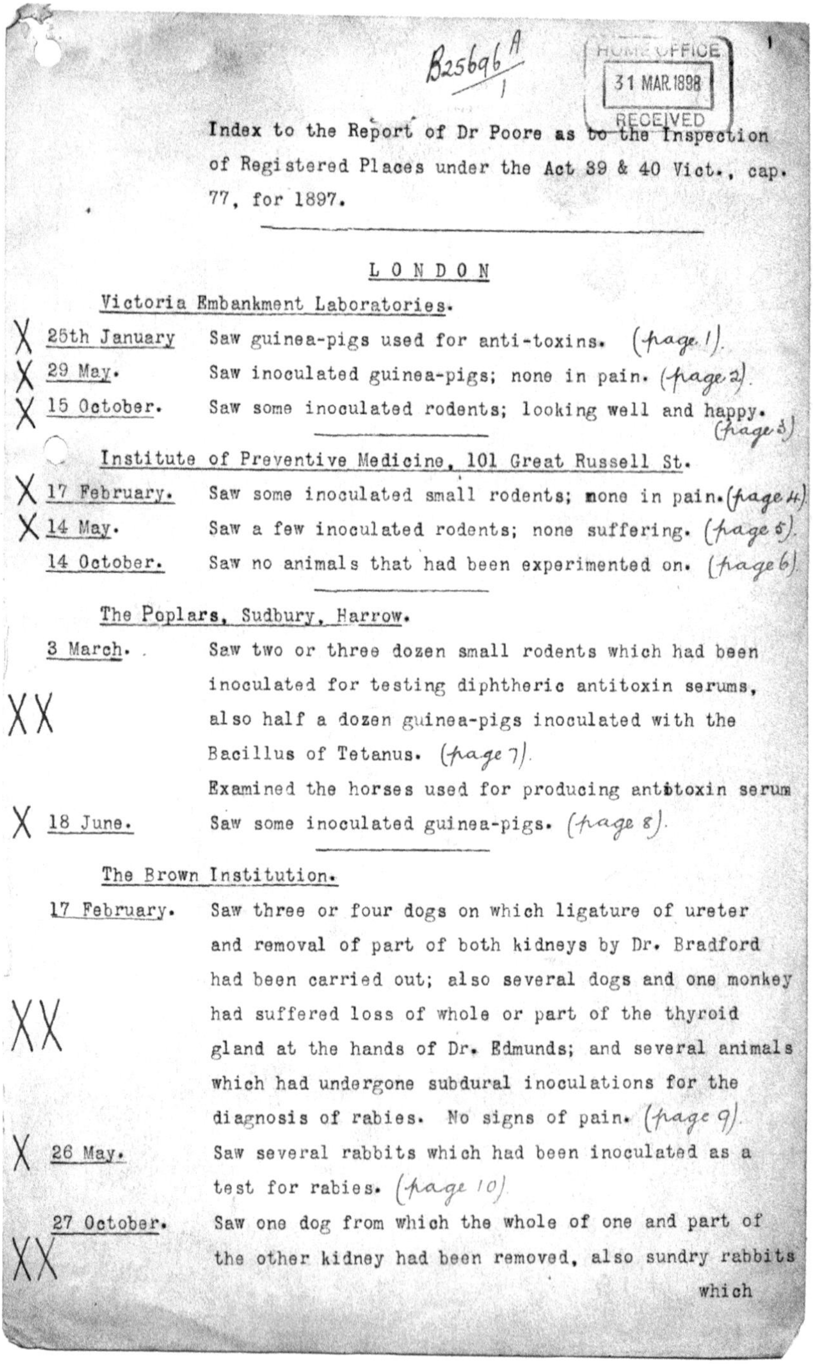

B25696 A
1

HOME OFFICE
31 MAR.1898
RECEIVED

Index to the Report of Dr Poore as to the Inspection
of Registered Places under the Act 39 & 40 Vict., cap.
77, for 1897.

L O N D O N

Victoria Embankment Laboratories.

X 25th January   Saw guinea-pigs used for anti-toxins. (page 1).
X 29 May.   Saw inoculated guinea-pigs; none in pain. (page 2).
X 15 October.   Saw some inoculated rodents; looking well and happy.
                                                              (page 3)

Institute of Preventive Medicine, 101 Great Russell St.

X 17 February.   Saw some inoculated small rodents; none in pain. (page 4)
X 14 May.   Saw a few inoculated rodents; none suffering. (page 5).
14 October.   Saw no animals that had been experimented on. (page 6)

The Poplars, Sudbury, Harrow.

3 March. .   Saw two or three dozen small rodents which had been
                  inoculated for testing diphtheric antitoxin serums,
XX   also half a dozen guinea-pigs inoculated with the
                  Bacillus of Tetanus. (page 7).
                  Examined the horses used for producing antitoxin serum
X 18 June.   Saw some inoculated guinea-pigs. (page 8).

The Brown Institution.

17 February.   Saw three or four dogs on which ligature of ureter
                  and removal of part of both kidneys by Dr. Bradford
                  had been carried out; also several dogs and one monkey
XX   had suffered loss of whole or part of the thyroid
                  gland at the hands of Dr. Edmunds; and several animals
                  which had undergone subdural inoculations for the
                  diagnosis of rabies. No signs of pain. (page 9).
X 26 May.   Saw several rabbits which had been inoculated as a
                  test for rabies. (page 10).
27 October.   Saw one dog from which the whole of one and part of
XX   the other kidney had been removed, also sundry rabbits
                                                              which

FIGURE 5 / Inspection of Registered Places in Great Britain, 1897. Dr. Poore's and Sir James Russell's reports, 1898, TNA, HO 144/419/B25696A.

'from time to time' in the Act have been interpreted in the Home Office to mean once a year," on another occasion, Matthews claimed that "two or more visits a year were expected to be made by the inspector" if experiments were being carried out.[63] When Russell was appointed assistant inspector, he was instructed to visit registered places about three times a year.[64] Poore visited some places only once a year, such as Wakefield and Wadsley, probably because the journey to these places was particularly burdensome, coupled with the fact that no experiments were being conducted there when he did visit. He visited other sites more often. For example, he inspected the Pathological Laboratory at London University College four times during 1894 and the same number of times in 1895. This is not surprising given that Poore was on the hospital staff at University College Hospital. Additionally, like many of London's registered places, this site was close to his residence on Wimpole Street, which might also have made it an attractive location for recurrent visits. These visits were recorded and reported back to the Home Office, documenting the law's encounter with real animals.

"There Can Be No Pain": Inspectors Meet Animals

In their visits to laboratories, inspectors attempted to evaluate whether the animals they saw—while undergoing an experiment or recovering from one—suffered pain and, if they did, to what degree. The historian Roy MacLeod observes that inspectorates in Britain "were acting as the 'eyes and ears' of government."[65] However, the vivisection inspectors were not only record keepers; they also made judgments about animals' well-being and proper conduct in the research laboratory. Enforcing the law was entangled with an experience of affect: the inspectors' observations exposed the workings of empathy and the way that individuals interpreted the otherwise abstract requirement to avoid painful experiments. What signs did the inspectors look for when trying to reflect upon the feelings or sensations of an animal?

Poore, Russell, and Thane, who were on duty in the last decade of the nineteenth century, left valuable handwritten and typed notebooks with their impressions from inspection visits. Their notes from 1894, 1895, 1897, and 1899

---

63. "Memo on Inspection"; Secretary of State to the Secretary, H.M. Treasury, December 12, 1890, HO 156/6, 46.

64. Appendix to First Report, 22.

65. Roy MacLeod, "Introduction," in *Government and Expertise: Specialists, Administrators and Professionals, 1860–1919*, ed. Roy MacLeod (Cambridge: Cambridge University Press, 2003), 1–26, 14.

provide a rare glimpse into how the Vivisection Act worked in practice.[66] The notebooks provide an informal record of the inspectors' practice and, together with memos and correspondences with the civil servants at the Home Office and with physiologists, afford a sense of the everyday life of the bureaucracy of empathy. The records reveal that the inspectors used various strategies to evaluate animal suffering. They examined some animals visually; at other times, they looked for changes in indicators of pain such as appetite or playfulness. On other occasions, an animal's physiological inability to feel pain, rather than visual indicators, convinced the inspectors that the animal was not suffering.

Annual reports about licensed experiments (sometimes called annual returns) carefully distinguished between painless and painful experiments. The annual reports to the Parliament included up until 1889 a count of "painful experiments." However, all the reports signed by Poore omitted an indicator of painful experiments, a fact that did not escape the notice of the Victoria Street Society.[67] In the first annual report Poore composed after his assignment as inspector, he stated that "the question of pain" did "hardly arise" in his inspections and added: "I am accustomed to examine all the animals minutely and individually, and I desire to state emphatically that it has never fallen to my lot to see a single animal which appeared to be in bodily pain."[68]

Poore encountered monkeys quite often, and he almost always took their energetic behavior as a sign of well-being. In 1894, Poore visited Charing Cross Hospital Medical School. "Saw some monkeys which were said to have been experimented on," he noted in his book, "saw nothing noticeable about them."[69] In a later visit, he saw a monkey that had had an operation performed on its cerebellum. Poore carefully scrutinized the body of the monkey and found "a slight deficiency of power in left hind leg." Four months later, Poore described one of the monkeys that he had seen before as "perfectly happy."[70] In January 1895, Poore visited Charing Cross Hospital Medical School again. Neither the physiologist Frederick Walker Mott nor his assistant, whom he had expected to see, was there to welcome him. The animals' room, however, was open, and Poore walked

---

66. Inspectors' visitation reports from 1896 and 1898 are missing.

67. Stephen Coleridge, An Open Letter to the Right Hon. Secretary of State, December 1900, HO 144/606/B31612.

68. Experiments on Living Animals: Copy of Report from the Inspectors Showing the Number of Experiments Performed on Living Animals during the Year 1890, under Licences Granted Under the Act 39 & 40 Vict. c. 77, Distinguishing Painless Experiments from Painful Experiments, Parliamentary Papers, 1890–1891 (266), 4.

69. Poore's Report on 1894, 31.

70. Poore's Report on 1894, 32, 33.

in to examine the animals. A suggestion of familiarity appears in his remark that the same monkey he had seen before was "as lively and friendly as usual."[71]

Poore consistently wrote positive reports, but his records reveal a more complex reality of mutilated animals. When Poore visited St. Thomas's Hospital in 1894, he saw two monkeys whose posterior nerve roots had been divided by Charles Sherrington. They were probably rhesus macaques on which Sherrington was exploring questions of motion and reflex action.[72] The monkeys had lost sensation or useful action in one of their hind limbs, and yet Poore could find no signs of suffering: "The animals were chattering and lively and certainly in no pain." On another visit to St. Thomas's Hospital, he saw a monkey that had some of its posterior nerve roots divided. He reported that neither the monkey nor several cats whose thyroid glands were removed seemed to be suffering. A few months later at the same institution he saw a monkey that "had its 'third nerve' divided. Its left eyelid drooped a little but as soon as its cage door opened it caught hold of the window rope and ran up it as only a monkey can."[73]

Poore rarely recorded inactive or frail monkeys. One of these encounters took place when he visited the Neuropathological Laboratory at King's College. Among all registered places, this facility was the hardest to get access to. Poore's earlier attempts to visit the site in May 1894 had failed, since the laboratory was locked.[74] Finally, in October 1894, he was able to enter the place. The single licensee associated with the place at that time was David Ferrier, who had been prosecuted a decade earlier by the Victoria Street Society for violating the Act; the charges against him were dismissed by the Bow Street Court (see chapter 2).[75] Poore encountered four monkeys, one of which had been trephined (the drilling of a hole in the skull) three days prior. It was sitting in front of a lit fireplace. Poore looked at the monkey and concluded that it "afforded no evidence whatever of pain or discomfort." He visited again the following year and saw a monkey that had had portions of its brain removed: "The animal was lively and active," though he noticed some problem in the hind legs. A few months later in the same place he saw a monkey that had had the motor cortex of the brain removed, "but it was quite nimble and in no suffering."[76]

---

71. George Poore, "Dr. Poore's Report of Inspection of Registered Places in 1895," 1896, HO 144/383/B19846A, 7 (hereinafter: Poore's Report on 1895).

72. F. W. Mott and C. S. Sherrington, "Experiments upon the Influence of Sensory Nerves upon Movement and Nutrition of the Limbs. Preliminary Communication," *Proceedings of the Royal Society of London* 57 (January 1, 1894): 481–88, 481.

73. Poore's Report on 1894, 25, 26, 27.

74. Poore's Report on 1894, 16.

75. Experiments on Living Animals, 1894, H.C. 103, 13.

76. Poore's Report on 1895, 18, 20.

Unpleasant sights were almost always absent from Poore's reports. The descriptions of the animals he saw were never harsher than, for example, a comment that a cat whose thyroid gland had been removed ten days earlier was "perhaps slightly sluggish in its movements." When he visited the Pathological Laboratory at London University College in January 1894, Poore saw a female dog from which half of the cerebellum had been removed some months ago. She was "happy and very proud of a puppy, the last of a litter of healthy pups she had recently brought forth." This kind of description was typical for Poore. When he saw a rabbit at University College in Liverpool, in which the sciatic nerve had been divided, he commented that it was "able to go about quite naturally and was in no discomfort."[77]

In light of the Vivisection Act, late nineteenth-century physiologists tried to regain authority over controversial issues such as vivisection by redefining the matter of dispute. In addition to replying to antivivisection claims by emphasizing the moral value of animal experimentation, they reiterated that animal pain is a consequence of physiological functions and hence in their realm of expertise. Identifying pain was thus a matter of expertise; in order for their evaluations of animal pain to be accepted as valid, inspectors had to be seen as scientifically informed authorities. Poore's employment of a scientific framing of animal pain was demonstrated in a visit to Brown Institution in London in October 1894, where he performed his own little experiment in sensation: "On entering I found a horse (a patient) which had . . . a severe wound on its right upper eyelid . . . The animal was standing quite quietly and apparently completely unconscious of its injury. I moved with my finger over . . . the wounded eyelid and the animal wouldn't move or withdraw. I mention this in relation to the question as to the power of animals to feel pain."[78]

Similarly, Assistant Inspector Russell's visitation notes were composed of short descriptions of the animals he had seen, followed by his almost-always positive impressions of their well-being. In February 1895, in the animal room of the University College of Newcastle, he found two monkeys, one of which was a pet belonging to chief servant Warwick and was therefore "not under the Act." The other monkey, named Congo, had had its thyroid removed by the physiologist George Murray long ago but showed no symptoms. Russell noted that in a previous visit, Congo was recovering from severe burns due to an accident, but this incident escaped his earlier reports and there is no indication that he reported the burn to Poore or to the Home Office. By then Russell thought Congo was

77. Poore's Report on 1894, 4, 6, 19.
78. Poore's Report on 1894, 41.

"quite well and very vigorous and tame. He escaped from his cage lately and showed great ingenuity in mischief." But Congo's playfulness did not last long. In August 1895, Russell was informed that the monkeys were infected with tuberculosis and had therefore been chloroformed to death. Russell noted how "the servant much lamented the loss of 'Congo' and 'Jenny' who followed him about the house as pets." Russell was taken by the analogy, and in December 1895 he remarked: "the animals at Newcastle are extremely well housed and treated as if they were pets in a private house."[79]

Two years later, in late 1897, Russell came for a visit, and Warwick showed him one monkey from which Murray had removed part of the thyroid. The monkey was eating well but was suffering from diarrhea. Originally, there had been two monkeys, probably replacing Jenny and Congo, but "one escaped in a mysterious manner when the under attendant opened its cage. It must be dead and a diligent search in the ventilation flues and all over the premises failed to discover the body."[80]

In the same year, Russell learned about another tragedy from Sherrington, who by then was doing research in the Physiological Department at the University of Liverpool: Six monkeys had been poisoned by escaping water gas, leaving Sherrington with only one big "untouched" monkey.[81] What Sherrington felt about this loss is hard to tell as he was, according to the historian Roger Smith, a reserved man "characteristic of his social world and its values."[82] The event however haunted the laboratory staff and during a visit to the place in 1899, a servant named Cox assured Russell that "by maintaining a steady temperature of 75 degrees monkeys will live in his care with very little risk of loss by death."[83]

Almost all animals held by Sherrington then were "more or less paralyzed in parts of the body," but Russell emphasized they were "sleek, fat and in good spirits." Two monkeys who had had a nerve root cut looked "very well and cheerful."[84] Years later, and after seeing some cases in which the removal of the

---

79. Russell's Report, in Dr. Poore's Report on Visitation of Registered Places in 1895, 1896, HO 144/383/B19846A, 13 (hereinafter: Russell's Report 1895).

80. Russell's Report in Inspection of Registered Places in Great Britain 1897, Dr. Poore's and Sir James Russell's Reports, 1898, HO 144/419/B25696A, 12 (hereinafter: Russell's Report on 1897).

81. This was the first time Russell inspected in Liverpool, which until 1897 was under the responsibility of Poore. According to Stella Butler, Sherrington carried out in Liverpool his most important neurophysiological work and laid the foundation for the modern discipline. Butler, "Centers and Peripheries," 477. See also Russell's Report on 1897, 18.

82. Roger Smith, "The Embodiment of Value: C. S. Sherrington and the Cultivation of Science," *British Journal for the History of Science* 33, no. 3 (September 1, 2000): 289.

83. Russell's Report in Reports of visits to Registered Places for year 1899, 1900, HO 144/451/B30824, 24 (hereinafter: Russell's Report on 1899).

84. Russell's Report on 1899, 27.

thyroid induced pain, Russell recalled those monkeys: "the first monkeys I saw with the thyroid removed were treated for a couple of years, under my observation, from time to time, with injection of thyroid extract, and that may have accounted for their not showing any symptoms. They were extremely lively and amusing, and evidently nothing had troubled them."[85]

The inspectors did not leave behind detailed criteria for judging the condition of animals, and they probably did not have any. Yet Russell explained that a loss of appetite was "the first and most valuable index that we have for the condition of feeling."[86] Reflecting back on his sixteen years of duty, Russell described his guidelines for identifying pain:

> My own private opinion is that whatever would effect the appetite or raise the temperature of an animal must be held to give some degree of pain, or of uneasiness amounting officially to pain. In some cases practically the only indication that the animal is suffering from anything is to find that the temperature is somewhat raised; or that the animal is losing flesh. I think want of appetite and loss of flesh are even better guides in the lower animals than they are in man. I do not mean to say that in every case of that sort there is severe pain. In my *précis* I have stated that it is a pain comparable to that felt by a human being suffering from diseases which would cause similar effects.[87]

Thane replaced Poore in 1899 and his notes echoed Poore's style and impressions. He even saw two monkeys sitting comfortably before the fireplace in the Neuropathological Department at King's College.[88] Thane's records demonstrate how the inspector was shaped to be both alert to animal pain and doubtful about its existence. Thane described his routine of visitation as the following: "I often inspect their registers, and I make a careful examination of the animals in the laboratory or in the animal houses, both those 'in stock,' and those which are under experiment; and the condition of the latter especially is noted and recorded."[89] Thane was confident when declaring that animals expressed "no indication" of pain, or showed "no signs" of suffering. But he was very cautious and unsure when there were indications or signs of thereof. When describing his visit to the Pathological Department at Cambridge he noted a couple of

---

85. Appendix to First Report, 23.
86. Appendix to First Report, 25.
87. Appendix to First Report, 23.
88. Thane's Report *in* Reports of visits to Registered Places for 1899, 1900, HO 144/451/B30824A (hereinafter: Thane's Report on 1899), March 6, 1899 (no page numbers).
89. G. D. Thane, The Royal Commission on Vivisection: Précis of Evidence, HO 144/4.

rabbits that were "only quiet and lethargic, did not show any signs of suffering."[90] A monkey in the Pathological Chemistry Department at University College in London had "perhaps" a little issue with one limb but Thane "could not be sure of it."[91] And in a visit to the Royal Veterinary College, "all the animals appeared perfectly well, with the doubtful exception of a rabbit . . . there is a considerable dwelling at the place of injection and there are some enlarged glands and the animal appeared to me to be somewhat wasted," however Thane reassured that "there is certainly no tenderness in the dwelling and so far as I could judge no pain, or even uneasiness."[92] In the Poplars in Sudbury Thane found an ill guinea pig that was "trembling and showed weakness" yet he "could not see any evidence of pain."[93]

Murray's Rabbit

The inspectors' reports conveyed their voices, and the archives contain only the descriptions of the experimental animals as the inspectors saw them. For the most part, to elicit perspectives other than the ones they provided, we can only try to read the inspectors' notes "against the grain" in order to deduce what information may have been omitted from their observations. The historian Keith Thomas describes it as "not so much for what the author meant to say as for what the text incidentally or unintentionally reveals."[94]

The following event offers an opportunity to compare Russell's point of view with another revealing source. In reports of visits during 1894, Russell mentioned two rabbits belonging to the physiologist George Murray, which he had seen before. Their thyroids had been removed long ago, but they were "well and lively."[95] In a later visit, Russell noted that "a rabbit which had the thyroid removed by Dr Murray two years ago was very fat."[96] But the physiologist's portrayal of the experiments reported by Russell evoked a very different image of the animals he used. In an 1896 *British Medical Journal* publication, Murray described the February 1893 removal of the thyroid gland from a black and white doe rabbit: "A month after the operation the rabbit had become rather dull and

---

90. Thane's Report on 1899, December 8, 1899.
91. Thane's Report on 1899, August 3, 1899.
92. Thane's Report on 1899, July 28, 1899.
93. Thane's Report on 1899, August 16, 1899.
94. Keith Thomas, June 10, 2010, https://www.lrb.co.uk/the-paper/v32/n11/keith-thomas/diary.
95. Russell's Report on 1894, 10.
96. Russell's Report on 1895, 3.

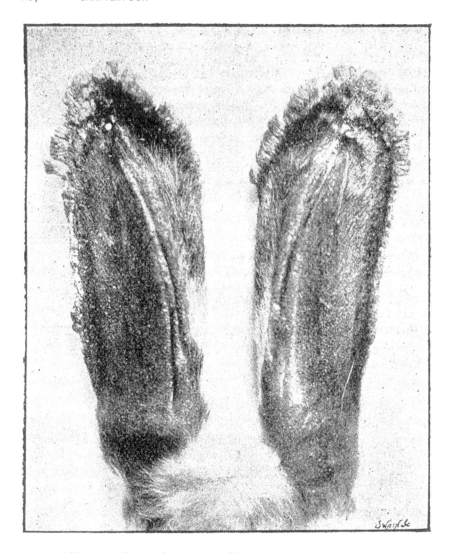

FIGURE 6 / An image showing the condition of the ears of a rabbit, which was probably seen by inspector James Russell in one of his visits to George Murray's Laboratory. Murray, *Effects of Thyroidectomy*, 205.

inactive, but it could be easily roused to active movements. A week later the appetite diminished considerably. . . . Then during February 1894, a good deal of the hair on the ears, and nose was shed. . . . The rabbit began to suffer from uterine haemorrhage and so it was killed."[97]

97. George R. Murray, "Some Effects of Thyroidectomy in Lower Animals," *British Medical Journal* 1, no. 1830 (January 25, 1896): 205.

The dates of this experiment and Russell's visits coincide, and chances are high that Russell saw the same rabbit that Murray had written about. Did Russell not see the cracks on the animal's skin? Did he not notice that it was "generally found sitting still with the eye half closed?"[98] In the same publication, Murray also remarked that monkeys after thyroidectomy suffer from similar symptoms of lethargy, swelling, reduced appetite, loss of hair, dryness of skin, and low temperatures. Did Russell see all this and yet decide to omit it from his report? Or did he think that these symptoms did not constitute suffering under the Vivisection Act? It is evident that Poore and Russell were reluctant to record indications of animal pain. As Murray's essay on rabbits showed, the inspectors might neglect to report troubling situations or, worse, turn a blind eye to cases of possible suffering. To a large extent, their avoidance reflected the inspectors' inclination to favor the experimenters and to protect their colleagues' reputations. But their caution with reporting about animals in pain also indicated their caution about asserting their views relating to issues in a field of knowledge that was increasingly ascribed to the realm of science.[99]

Animal suffering, if the inspectors acknowledged it at all, was seldom included in their reports, but they translated it to a set of procedural requirements. As the next section will show, the inspectors used their power to the fullest when confronted with mundane bureaucratic breaches of the Vivisection Act, such as the lack of a proper license or an expired certificate. The inspectors mostly constructed pain as an object of regulation not by a direct identification and handling of the suffering body but, rather, through the creation of various other means—such as meticulous enforcement of procedural requirements, the education of physiologists about the Act, and intervention in the design of the experimental space—in the name of animals' well-being. Through this bureaucratic oversight, the inspectors facilitated the incorporation of the standards of behavior prescribed by the Vivisection Act.

Infringements

Visits to laboratories were a central method for revealing infringements relating to experiments without the required authorization. Inspectors would ask experimenters to produce certificates relating to the animals they had seen in the laboratory. In cases where the experiments did not require certificates, the inspectors

---

98. Murray, "Some Effects of Thyroidectomy in Lower Animals," 205.
99. Otniel E. Dror, "The Affect of Experiment: The Turn to Emotions in Anglo-American Physiology, 19001940," *Isis* 90, no. 2 (June 1, 1999): 205–37.

would examine the physiologists' licenses to make sure that they were up to date. W. P. Byrne, a principal clerk in the Home Office, estimated that during the first three decades of the Act some sixty "cases of contravention" of the Act had come to the notice of the Home Secretary, "the great bulk of them being of a trifling character and calling for no step beyond a warning or a rebuke."[100] Reports on infringements could lead to the termination of a license or to substantial restrictions on the ability to conduct experiments on animals. In reality their most apparent effect was only to cause some delays in the work of the physiologists.

In 1894, Poore included two infringements of the Vivisection Act in his annual report to Parliament. The licensees, he explained "carelessly exceeded the powers given to them by their certificates."[101] One of these infringements was discovered when Poore visited the Charing Cross Hospital Medical School, where he saw a dog on which neuropathologist Frederick Walker Mott had made a central incision in the spinal cord. Mott did not have the proper certificate for this experiment, "but this was reported at the time and settled."[102]

In mid-November 1895, Poore visited the Pathological Laboratory at the Army Medical School in Netley, where he met the bacteriologist and immunologist Almroth E. Wright. He had observed a rabbit dying "perfectly tranquilly" as a result of subdural inoculation of hydrophobia (rabies) ten days earlier. However, allowing an animal to recover from the influence of anesthetic after an experiment required a special certificate; Poore ascertained that Wright did not hold the appropriate certificate to cover the procedure and informed the Home Office.[103] Wright argued that he had obtained the correct certification but failed to include the specific operation on the rabbit. Home Secretary Ridley accepted Wright's explanation, and, following an arrangement made between Poore and Wright, he requested that Wright submit a corrected certificate.[104]

Not all experimenters were as lucky as Mott and Wright. On January 11, 1895, Poore visited the examination rooms of the Royal Colleges of Physicians and Surgeons in London. In the animal room, among various guinea pigs and rabbits, one cat caught his attention. Experimenter Fenwick had created a gastric fistula in the cat a few days previously, but the cat seemed to Poore to be "unconscious of its condition." Poore had the cat taken out of its cage and noted

---

100. Memorandum of Evidence to be Given by Mr. W. P. Byrne, C.B., A Principal Clerk in the Office of H.M. Secretary of State for the Home Department, HO 114/4.

101. Experiments on Living Animals, 1894, H.C. 103, 41.

102. Poore's Report on 1894, 7, 32.

103. Poore's Report on 1895, 43.

104. Letter on behalf of the Secretary of State to Professor Wright, December 7, 1895, HO 156/9, 524.

that it was "quiet and purring and rubbed its head against my closed hand as often the manner of cats when we think they are pleased." However, four days later, Poore returned and informed the director of the institution, German S. Woodhead, that Fenwick had performed the experiments without the certificates that were required: first, to experiment without anesthesia and, second, to keep an animal alive after the experiment. Poore and the director went together into the animal room to watch the cat, where once again the animal seemed to Poore to be "unconscious of its condition." When it was taken out of its cage, the cat wandered around the room purring and rubbing up against Poore, "affording no evidence whatever of discomfort."[105] Although Poore emphasized in his notes that the cat was in good condition, he reported the infringement of the Act relating to the certificates to the Home Secretary.

Two weeks later, a letter from Home Secretary Asquith was sent to the president of the Royal College of Physicians stating that "the Secretary of State having carefully considered all the circumstances of the case had decided that Dr. Fenwick's license . . . must be revoked and the Certificate A3 held by him disallowed."[106] Letters were also sent to Fenwick and to the secretary of the Association for the Advancement of Medicine by Research.[107] Poore was informed about the situation that resulted from the information that he provided and for which he was, to a large extent, responsible.[108] Poore revealed an additional infringement of the Vivisection Act later in the same month. Visiting the Brown Institution, he noted that Sherrington did not have the certification required for the experiments he performed on cats and reported the matter to the Home Secretary.[109] In April 1895, the Home Office sent a long letter to Sherrington. While Home Secretary Asquith acknowledged that Sherrington had performed several experiments without holding the requisite certificates, he was ready to accept that the infringements were done "through forgetfulness." However, he stressed the "necessity of an invariable compliance with the provisions of the Act" and concluded, with regret, "that he feels bound to mark his sense of [Sherrington's] neglect to observe the law" by postponing for a month the further consideration

---

105. Poore's Report on 1895, 10, 11.

106. Letter on behalf of the Secretary of State to the President of the Royal College of Physicians, January 31, 1895, HO 156/9, 193.

107. Letter on behalf of the Secretary of State to the Secretary of the Association for the Advancement of Medicine by Research, January 31, 1895, HO 156/9, 192; Letter on behalf of the Secretary of State to Fenwick, January 31, 1895, HO 156/9, 190.

108. Letter on behalf of the Secretary of State to George W. Poore, January 31, 1895, HO 156/9, 194.

109. Poore's Report on 1895, 4.

of certificates submitted by Sherrington.[110] On May 11, 1895, Kenelm Digby, permanent undersecretary, informed Sherrington that his license was renewed and his certificates allowed.[111]

This example shows that, contrary to the claim made by the historian Richard French, the Vivisection Act's enforcement was not directed solely toward the weaker and less influential members of the experimental world, as Sherrington had been penalized despite his position as the general secretary of the International Physiological Society.[112] Other examples of civil servants confronting senior scientists exist. For instance, the Home Office denied a certificate to the established physiologist Gerald Yeo, who complained to his colleagues that his work had been obstructed by the Home Office (see chapter 3).

Despite the infringements made by Wright, Fenwick, and Sherrington in 1895, when Poore submitted his annual report to Parliament, he mentioned only two cases of irregularities. (He probably omitted from his count Wright's case, which did not entail punitive measures.) Poore maintained that "the licensees have, as usual, manifested strict loyalty to the letter and spirit of the Act."[113] But some, such as MP MacNeill, were unsatisfied with the obscure description of the irregularities provided by Poore. MacNeill asked Home Secretary Ridley for details about the irregularities mentioned in the annual return for 1895, and about the proceedings that had been taken against the violators. Ridley, who was newly appointed and was not the one to make the decisions in those cases, provided some information about Fenwick's and Sherrington's cases, without revealing their name. He rejected MacNeill's accusation of "diabolical cruelty," but added that the two cases had "a great difference between them," implying that Fenwick's lack of certificate was worse than Sherrington's omission to renew his.[114]

In the same protective tone, when Poore reported in 1897 seven cases of licensees "inadvertently overstepping their powers under the Act," he noted that "the facts were recorded in good faith by the licensees themselves, and in each case they expressed their great regret at the inadvertence."[115] Inspector Russell expressed the same confidence in the experimenters when he wrote to the undersecretary of state that he found that "many licensees had difficulty in

---

110. Kenelm Digby to Charles Sherrington, April 3, 1895, HO 156/9, 259.

111. Kenelm Digby to Charles Sherrington, May 11, 1895, HO 156/9, 316.

112. Cf. French, *Antivivisection and Medical Science*, 208.

113. Copy of Report from the Inspectors Showing the Number of Experiments Performed on Living Animals during the Year 1895, Parliamentary Papers, 1896 (330), 3.

114. 45 Parl. Deb. (4th ser.) (1897) 1288.

115. Experiments on Living Animals, 1898, H.C. 215, 695.

understanding the Act," but he also "saw nothing to indicate that any of them had failed to comply with it or desired to evade its provisions."[116]

The annual reports composed by the inspectors and presented to the House of Commons by the home secretary routinely portrayed the experimenters in a positive, cooperative light, and downplayed their misdeeds. Higher officials echoed this message. Home Secretary Ridley claimed before the House of Commons that all cases of infringements during the years 1894–97 had risen "from inadvertence" and were not "fitting cases for prosecution." Ridley declined a request to publish the names of the persons who had performed illegal experiments, and explained that he did not think it necessary or right "where the illegality has not been of so serious a nature as to call for the institution of legal proceedings."[117]

Too Bold and Too Faint: The Critique of the Inspectorate

The official reports did not detail important aspects of the communications between inspectors and experimenters behind the scenes. The inspectors' visitation reports were confidential and so was the frequent correspondence between the home secretary and the inspectors. The general public, including members of antivivisection societies, could only learn about the administration of the Vivisection Act from the published annual reports and the sporadic references to the subject made by the home secretary during parliamentary debates that found their way to the newspapers.

The way in which inspection was carried out was so obscure that it led antivivisection protestors to doubt that any inspection was taking place. A contributor to the *Zoophilist* complained that while inspectors of schools were so meticulous as to record the beginning and ending times of each of their school visits, "there is not even the pretense of registering the so-called inspectors' visits." None of the inspectors, he contended, "has condescended to afford the nation which had paid their salary the slightest information as to when or where—if ever—they have inspected a physiological laboratory at all."[118] A reviewer of the

---

116. James Russell to the Undersecretary of State, March 5, 1900, HO 144/451/B30824. Inspectors in other fields also felt they had to educate and simplify the Act for people (Pellew, "The Home Office and the Explosives Act of 1875," 190–91).

117. 62 Parl. Deb. (4th ser.) (1898) 835–36.

118. "Notes and Notices," *Zoophilist* 7, no. 5 (September 1, 1887): 74.

1893 annual report frustratingly commented that "it would have been much more to the purpose if Dr. Poore had said something more of the nature of the 'experimental work' itself, something of the fate of the animal on which it is carried out; but this is just what he avoids."[119]

A constant complaint was that the inspectors' annual returns were based more on the scientists' own accounts of their laboratory conduct than on the inspectors' field reports. As one MP put it, "under no other system of inspection would it be allowed that the very individuals whom it was desired to keep under inspection should make their Reports unchecked to the House. An Inspector of Mines did not report on the Reports furnished to him by the owners or the workers in the mines, nor would an Inspector of Factories report on the ipse dixit of the owners or representatives of factories."[120]

The *Zoophilist* made the accusation that the annual report was nothing but "anaesthesic to public opinion" and that "the inspector flounders deeper and deeper into the morass of insincerity as he strives his hardest to screen his clients, the vivisectors, from the public odium and condemnation they so richly deserve."[121] A year later MP MacNeill claimed that "the inspections of the vivisection laboratories were absolutely worthless" and complained that the home secretary "had refused to give a return showing the number of times the inspectors visited the laboratories and describing what they saw."[122] But his call was never answered; inspectors did not describe what they saw and published only their final judgments on pain.

Since the Home Office did not disclose to the general public much of the routine administration of the Act, critics of the Act concluded that experimenters dictated what the inspectors wrote in their reports. When the *Derby Daily Telegraph and Reporter* summarized one of the annual reports to its readers, it added that "the reporting officer, however, depends upon the assurance of the experimenters that in none of the other experimenters under the certificates was any appreciable suffering inflicted."[123] Similarly, the *Zoophilist* published that the inspector "only knows what the vivisectors tell him."[124] A reviewer of the 1897 annual report was particularly vexed by what he understood to be a statement

---

119. Legalized Vivisection in 1893," Zoophilist 14, no. 2 (June 1, 1894): 24.

120. 17 Parl. Deb. (4th ser.) (1893) 331–32.

121. "The Inspector's Report and Vivisectors' Returns for 1896," *Zoophilist* 17 no. 3 (July 1,1897): 47.

122. 52 Parl. Deb. (4th ser.) (1897) col. 373.

123. "Report on Vivisection," Derby Daily Telegraph and Reporter 277 (June 14, 1880): 3.

124. "The Inspector's Report and Vivisectors' Returns for 1896," *Zoophilist* 17 no. 3 (July 1,1897): 47.

solely based on information dependent on the goodwill of inspectors: "this shows that the vivisectors concerned were their own inspectors; that there were breaches of the law, but that the inspectors would not have known of them if the delinquents had not confessed their offences."[125] And on another occasion, the chaplain of Lucas's Hospital in Wokingham, Joseph Stratton, claimed that it was "highly probable that he [the inspector] knows very little as to what really goes on in the torture-chambers up and down the country."[126]

While antivivisection advocates complained that the inspectorate was inactive, physiologists felt it was overactive, as demonstrated in the 1881 report of the Physiological Society, which testified to the impact the Act had on vivisection practitioners (see chapter 2).[127]What the Physiological Society described in its report are scientists who feel discouraged by the threat of administrative action, and their realization that they must find ways to cope with the new legal-moral order at the same time as trying to adjust their needs to it. The Home Office's published records purposely omitted the disciplinary actions taken against experimenters, such as the suspension of a license or the disallowance of a certificate, from the public eye. But other records show that, even though the inspectors' reports to the home secretary did not lead to any prosecution in the period under study, they were effective in making the experimenters adhere to the Act's requirements. The inspectors' impact on the experimental setting is also demonstrated in their involvement in its design.

Ingenious Cages: The Registration Requirement

The inspectors did not openly advocate administrative discretion but exercised it to the fullest in the registration of places for experiments.[128] Section 7 of the Vivisection Act authorized the home secretary to require the registration of places for the performance of experiments, as a condition for granting an experimentation license. The Home Office made registration mandatory, and

125. "The Inspectors' Report and Vivisectors' Return for the Year 1897," *Zoophilist* 18, no. 3 (July 1898): 54.

126. Jos. Stratton, "Inspectors under the Vivisection Act," Zoophilist 13 (October 1894): 91.

127. *Report of a Committee Appointed by the Physiological Society* (in pursuance of the following resolution passed on October 15, as amended and adopted by the Society at its meeting of December 8, 1881), n.d., file SA/RDS/A/3, Wellcome Archives, London.

128. Cf. Jessica Wang's claim that New Deal lawyers "built the case for expertise and administrative discretion as means for the flexibility necessary to manage the corporate political economy." Jessica Wang, "Imagining the Administrative State: Legal Pragmatism, Securities Regulation, and New Deal Liberalism," *Journal of Policy History* 17, no. 3 (2005): 271.

the inspectors were in charge of processing the registration forms. The inspectors saw the experimental environment as an indication of the degree of attention being given to the well-being of animals, and, with this justification, they gained an increasing influence over the design of laboratories and animal rooms. They examined applications for the registration of such buildings, visited the laboratories nominated for such registration, and submitted their recommendations to the home secretary. Without the inspectors' approval, an experimental project could not have been accomplished.

Most notably, the Act contributed to the declining number of private experimental laboratories.[129] The General Medical Council was opposed to the registration provision when it first appeared in the bill. The council maintained that only spaces used for instructional purposes should be required to register. It argued that limiting experiments to registered places "would not only tend seriously to obstruct genuine scientific inquiry, but would also prove impossible in practice." The council further claimed that the licenses and certificates required from the individual physiologists were enough to ensure the proper conduct of experimenters, and, therefore, "private researches, which in competent hands might prove of the highest value to mankind . . . ought by no means to be prohibited."[130] The Parliamentary Bills Committee of the British Medical Association also recommended changing the bill so that "a licensed person may register any convenient place for the purpose of experiment, and that such register shall be confidential document," accessible only to the secretary of state and appointed assistants.[131] The secrecy of the registration was requested to avoid antivivisection protests.

A similar dissatisfaction with the registration requirement was voiced in a memorandum signed by "Teachers of physiology in England, Scotland and Ireland," among whom were William Sharpey, William B. Carpenter, Michael Foster, Gerald Yeo, and Arthur Gamgee. These leading physiologists were willing to accept the restriction of teaching demonstrations to registered laboratories and in principle also of research experiments. But they claimed that flexibility was necessary when investigations had to be undertaken at short notice or at a distance from laboratories, in experimenters' own houses or elsewhere.[132] Despite

---

129. On the domestic and public domains of the laboratory, see Graeme Gooday, "Placing or Replacing the Laboratory in the History of Science?," *Isis* 99, no. 4 (December 2008): 783–95.

130. General Medical Council, "Memorial," 3–4.

131. Parliamentary Bills Committee of the British Medical Association, "The Cruelty to Animals Bill," n.d., file SA/RDS/A/3, Wellcome Archives.

132. Parliamentary Bills Committee of the British Medical Association, "The Cruelty to Animals Bill," 5.

this opposition, the Act allowed the Home Secretary to mandate the registration of all places used for animal experimentation.

The Home Office was inclined to disallow the registration of private residences as experimental spaces although the Act made no such restrictions. Home Secretary Matthews had adopted the policy in the 1880s following the advice of inspector Erichsen to prefer places to which "a certain publicity attaches."[133] Erichsen informed the home secretary in 1888 that "all the experiments are done more or less in public, that is to say, so secrecy is observed, no door is locked, there is free access to the laboratories or rooms in which they are performed."[134] Matthews reasoned that in public places there is a presence of persons "who understand the subject, and can see that the thing is done in accordance with the law." In places such as the Brown Institution, Owens College, and the University of Aberdeen, "there is always an intelligent and enlightened audience at hand, able and ready to check anything like abuse." Restricting experiments to public spaces, he asserted, was "one of the greatest and most valuable checks and safeguards that could be introduced."[135] Matthews's statement demonstrates again how the public act of witnessing produced something more than the scientific fact that the experiment sought to establish. Having a designated public space, inhabited by licensed experimenters and open to inspection, helped to create a cruelty-free fact, or in other words, lawful knowledge.

For this reason, Russell was perplexed by his visit to Glasgow in November 1890, when he met with the pathologist Joseph Coats, and discovered the peculiar arrangements Coats created for his animal subjects. After inoculating the animals in the Physiological Department, which was a registered place, he took the animals to a "small well lighted attic" in the Pathological Department, which was not registered under the Act. The Pathological Department was closer to Coats's home, which made it easier for him to attend to the animals.[136] Russell was concerned that Coats's arrangement contradicted the requirement to experiment only in a registered place. He consulted Poore and explained that Coats had difficulty leaving his animals in the registered place, speculating, "possibly his Pathological animals might not be very welcome in the Physiological laboratory."

Poore forwarded the inquiry to the Home Office, together with an additional case in which Bokenham (probably Thomas Jessop Bokenham) wished to remove inoculated rabbits from a registered place to a private estate outside of London.

133. 335 Parl. Deb (3rd ser.) (1889) 886.
134. Erichsen to Matthews, February 6, 1888, HO 144/299/B2697.
135. 35 Parl. Deb (3rd ser.) (1889) 886.
136. Russell to Poore, November 24, 1890, HO 144/322/B9399.

In regard to the latter, Poore commended that such a removal would be "most desirable" considering the comfort of the animals.[137] Although "as a general rule" the home secretary had refused to register a private residence, the Home Office clarified that an exception could be made: "it will be better for the animals and will at the same time help to make the experiment more trustworthy."[138] Registration of such place would be accepted by the Home Secretary, under the condition that the names of both places—the place of inoculation and the place of housing—should be inserted in the license.[139] Following this decision, annual returns published after 1890 included the statement that all licenses were restricted to the licensed places, "with the exception of inoculation experiments in places other than 'licensed places' with the object of studying outbreaks of disease among animals in remote districts."[140]

Coats was unsatisfied with the decision. Registering the laboratory could draw unwanted attention from critics of vivisection. He anticipated difficulties in convincing the managers to register the Pathological Laboratory as a place in which vivisection was practiced since "it might possibly affect subscriptions to the institution which depends upon the public."[141] In February 1891, the home secretary clarified that the removal of the animals to an unregistered place was impossible.[142]

Coats's concern demonstrated the chilling effect the registration requirement had on some institutions.[143] However, by the mid-1890s research institutions largely cooperated with the Act's requirement of registration. The initial hostile response of representative scientific organizations to the idea of registrations was based on the tradition of home laboratories, which was fading away also for other reasons. By the 1890s, most experiments were performed in university or hospital laboratories.[144] The attempts to keep experimentation in private places lawful turned irrelevant with the accelerated changes in the physiological

---

137. Poore to the Undersecretary of State, November 25, 1890, HO 144/322/B9399.

138. Copy of minute by Godfrey Lushington, November 28, 1890, HO 144/322/B9399.

139. Lushington to Poore, December 2, 1890, HO 156/6, 33.

140. Experiments on Living Animals: Copy of Report from the Inspectors Showing the Number of Experiments Performed on Living Animals During the Year 1890, under Licences Granted under the Act 39 & 40 Vict. c. 77, Distinguishing Painless Experiments from Painful Experiments, Parliamentary Papers, 1890–91 (266), 3.

141. Extract from Dr. Russell Report to Dr. Poore, December 17, 1890, HO 144/322/B9399.

142. Memorandum by Godfrey Lushington, February 27, 1891, HO 144/322/B9399.

143. French, *Antivivisection and Medical Science*, 283.

144. E. M. Tansey, "The Wellcome Physiological Research Laboratories 1894–1904: The Home Office, Pharmaceutical Firms, and Animal Experiments," *Medical History* 33, no. 1 (January 1989): 6.

profession, its incorporation into medical schools and hospitals and the establishment of physiological chairs in many leading institutions.

What criteria did the inspectors use to determine whether places were suitable for registration? Twice in 1899, James Russell declined to recommend places for registration "owing to defects of light and ventilation," though these animal houses "had been specially designed by architects of repute."[145] At the University of Glasgow, he found the heating system satisfactory but traced problems in the lighting as well as in the low ceiling that could hurt the ventilation. His host, Dr. Muir, was unwilling to accept Russell's decision, stressing that an eminent architect had designed the building. But Russell "could not give way." He explained to Muir that his experiments would require more than a hundred animals at a time "and that it was not desirable to sanction inferior animal houses in a new building to be registered for the first time." Russell and Muir finally agreed that, until an appropriate outside space could be provided, "one of the good rooms on the main flat should be used as an animal house."[146]

That year Russell also declined the registration application of the Medical School at the University of St. Andrews in Scotland. A few days after his visit to Glasgow, Russell met with a physiologist named Harris and an anatomist named Musgrove to inspect the premises at the Medical School. In contrast to many other establishments in which animals were housed in improvised facilities, the animal house in St. Andrews had been expressly built for this purpose. Yet Russell was dissatisfied with the space's ventilation and thought it would not be sufficient for many animals. He provided Harris with a letter for the architect and the building committee, outlining his recommendations to remedy the defects he had identified.[147] They made the required alterations in the plans and Russell approved the application.[148]

Inspectors' influence on the design of the experimental space was also demonstrated by Russell's involvement in the planning of an animal house and an experimental room in Worcester. In mid-1899, Russell met some members of the county council and together they inspected some spaces that they intended to register for experiments related to public health. Russell sent a full report regarding this visit to Poore since it was the first municipal laboratory he was asked to register.[149]

---

145. James Russell to the Undersecretary of State, March 5, 1900, HO 144/451/B30824.
146. Russell's Report on 1899, 11.
147. Russell's Report on 1899, 21.
148. James Russell to the undersecretary of state, March 5, 1900, HO 144/451/B30824.
149. Russell's Report on 1899, 32.

The inspectors' scrutiny of the experimental space extended beyond the registration process and included evaluating and mandating appropriate adjustments where necessary in the living environments of animals in physiological laboratories. Poore described the living conditions of animals in general terms: he thought that the laboratory at University College in Cardiff "was in every way fitted for the performance of experiments," and he found the Physiological Department at Cambridge "beautifully clean and the animals most comfortable."[150] Russell, perhaps due to his experience in public health administration, was more detailed in his explanations. In a visit to the Pathological Department at the University of Aberdeen, Russell was impressed by an "ingenious cage with a false floor for the sake of dryness," which accommodated four guinea pigs.[151] In the Materia Medica Department at the University of Aberdeen he saw "a comfortable box with plenty of food," as well as six dozen healthy frogs who were "kept in tanks with blocks of wood charcoal upon to climb," while the frogs in the University of Glasgow were kept in a ranarium and were covered with leaves. Other senses were also sometimes involved: Russell reported that the animals in the Materia Medica Department at the University of Edinburgh were very well cared for and noted that there was "no odour in the rooms."[152]

Cold was a considerable threat to the lives of animals in laboratories, and Russell often commented on the heating system in the animals' rooms. At the Physiological Department at the University of Edinburgh he visited a workshop warmed by hot pipes.[153] With the pathologist David James Hamilton he inspected a new animal house on the roof of the Pathological Department at the University of Aberdeen, while plumbers were fixing hot-water pipes. Russell believed that expenses would prove too high.[154] Others wrapped the animals with straw to keep them warm. At the Royal College of Physicians Laboratory in Edinburgh, six guinea pigs and eleven rats were packed in straw. But the winter of 1895 was harsh. The water pipes froze and not much was going on in the facility when Russell came for a visit. All the guinea pigs had died and the surviving animals were covered with straw.[155]

Rather than being "a mere formality once officially requested," each registration application was carefully considered by the inspectors, and their views had an important impact on the experimental space as experimenters

---

150. Poore's Report on 1894, 3, 2.
151. Russell's Report on Visitation of Registered Places during 1894 HO 144/370/B17451A, 8.
152. Russell's Report on 1895, 8, 3.
153. Russell's Report on 1894, 54.
154. Russell's Report on 1897, 9.
155. Russell's Report on 1895, 5, 6.

gradually incorporated the inspectors' guidelines in the design and operation of their research facilities.[156] Experimenters consulted Russell about appropriate arrangements for holding experimental subjects, and Russell testified that he was frequently asked for advice "regarding improved methods of construction and warming of houses for animals."[157]

By the turn of the twentieth century, registration was becoming a marker of rigorous research endeavor. For the pharmaceutical manufacturer Burroughs Wellcome and Company, registering laboratories for vivisection had become a matter of principle. It was essential in order to both accomplish its commercial plans of drug production and to assert its legitimacy as a research institution among the established physiological laboratories that were mainly associated with medical schools and hospitals. Bokenham, director at Wellcome Laboratories, explained that he was following Poore's advice when applying to register his place, "to guard against possible contingencies and to avoid the annoyance and loss of time which might conceivably be incurred in meeting unfounded charges" by antivivisection advocates.[158]

After inspecting the Wellcome Physiological Research Laboratories facility in Charlotte Street, Poore concluded that the place was fit and registration should be allowed, but not before the principal questions of registration of a commercial place and its consequences were thoroughly examined. An official application was therefore submitted only in 1900, after the laboratory was relocated to Brockwell Hall in South East London. Inspector Thane went to inspect the new place, and returned as impressed as Poore had been before him.[159] Wellcome's representatives had an informal meeting with Thane, where they conversed "as one scientific man to another."[160] Responding to the Home Office's concern that many other commercial companies would follow suit (detailed in chapter 3), Wellcome claimed that "inferior firms will not have the advantages of the facilities and equipment that we have."[161] At this stage, the established scientific institutions resisted the registration of Wellcome.[162] The Laboratories Committee of the Royal College of Physicians and the Royal College of Surgeons of England presented a few arguments against it, among which, the claim that experiments

156. French, *Antivivisection and Medical Science*, 283.

157. James Russell to the Undersecretary of State, March 5, 1900, HO 144/451/B30824.

158. T. J. Bokenham to the Home Secretary, September 7, 1896, HO 45/11092, 3.

159. Tansey, "Wellcome Physiological Research Laboratories," 8, 19.

160. Dowson to Wellcome, July 27, 1900, WF/180, WPRL (quoted in Tansey, *Wellcome Physiological Research Laboratories*, 25).

161. Notes re Petitions to Home Office, July 20, 1900, WA, WF/WPRL/01/02, 2.

162. Tansey, "Wellcome Physiological Research Laboratories," 23.

made for the purpose of trade "have not the same independent character as the purely scientific and human experiments . . . which are carried under the Act."[163]

The first WPRL application was rejected, but a second successful application was submitted in 1901. This time Wellcome had hired the legal services of Fletcher Moulton, who advised the company to include in its application as few experiments as possible "so that the Home Office should not be alert at the amount of work we intend to do at the Laboratories, for the main thing is to get the Laboratories registered."[164] The company also emphasized the eligibility of its experimental spaces, stating that the experiments would take place "in a separate room to be especially reserved for this class of work." It added that the room would be "admirably adapted for the purpose" and expressed its will to show the room to Home Office representatives.[165]

Early in the registration procedures and again during the examination of its 1901 application, the Home Office suggested that Henry Wellcome keep the production of the company's drugs in his facilities but conduct the related experimentation in laboratories that were already registered with established research institutions. Wellcome rejected this advice, complaining that it entailed an unnecessary burden on his staff. He also referred to the symbolic element in the registration process: "I desire to come openly under the existing law, and to have my Physiological Research Laboratories registered under the act controlling animal experiment" to be "regularly inspected like other laboratories of the same kind."[166]

The regulation of experimental spaces and their separation from scientists' private lives is part of the law's expanding oversight of spaces and behaviors. But, as Susan Silbey and Patricia Ewick note, the subjection of the experimental space to law had strengthened science's objective appeal and reasserted the authority of the knowledge it produced.[167] However, the regulation of animal experimentation shows that the law not only contributed to the authority of science but also eased public critique of its practices. The Vivisection Act also shaped core methods of medical research such as witnessing and made bureaucrats into actors in the coproduction of knowledge and social order.

---

163. P. H. Pye-Smith, Report, July 3, 1900, HO 45/11092.

164. A letter draft by SM, April 4, 1901, WA, WF/PRL/01/03; SM to Mr. Fabian, July 5, 1901, WA, WF/PRL/01/03.

165. A letter draft to the Undersecretary of State, May 21, 1901, WA, WF/PRL/01/03.

166. A letter draft to Home Secretary Matthew Ridley, February 15, 1900, WA, WF/WPRL/01/02, 2.

167. Susan Silbey and Patricia Ewick, "The Architecture of Authority: The Place of Law in the Space of Science," in *The Place of Law*, ed. Austin Sarat, Lawrence Douglas, and Martha Merrill Umphrey (Ann Arbor: University of Michigan Press, 2003), 75–108, 79.

Modest Witnesses for Cruelty-Free Facts

The inspectors were responsible for shaping the experimental space as an ethical environment, and their persona was designed to bear witness to the morality of its products. Vivisection inspectors acted as "modest witnesses," a term used by Steven Shapin and Simon Schaffer, who argue that modern experimental method necessitated a collective act of witnessing and a reporter who was accepted as a reliable provider of testimony to create an experimental fact.[168] Donna Haraway readdresses the figure of the modest witness, analyzes how witnesses gain their legitimacy and authority over the establishment of scientific facts, and explains that "in order for the modesty . . . to be visible, the man—the witness whose accounts mirror reality—must be invisible, that is, an inhabitant of the potent 'unmarked category,' which is constructed by extraordinary conventions of self-invisibility." The self-invisibility that the modest witness enjoys is modern, European, and masculine. It provides him with "the remarkable power to establish the facts. He bears witness; he is objective; he guarantees clarity and purity of objects."[169] In the case of animal experimentation, civil servants acted as reliable reporters; through the Vivisection Act's implementation, inspectors were figures of modest witnesses who enabled the creation of an ethical scientific fact.[170] The participation of inspectors as witnesses and reporters for the production of science was pushed to the extreme when they were asked to watch live experiments.

Some critics of vivisection demanded that inspectors be regularly present at experiments. For them, examining the unused animals, or those who had already recovered from vivisection, was not enough. Public pressure forced the Home Office to consider sending inspectors to watch live experiments. In 1890, inspector Erichsen doubted that the Home Office could appoint inspectors whose duties would be to attend and supervise the performance of experiments. Erichsen believed that "few if any gentlemen of independent position" would undertake a mission with such an "inquisitorial character." And, on a more practical note, Erichsen predicted a problem arising from the simultaneous scheduling of experiments. Home Secretary Matthews clarified at the same time that the Act did not require the personal supervision of inspectors over

---

168. Steven Shapin and Simon Schaffer, *Leviathan and the Air-Pump: Hobbes, Boyle, and the Experimental Life* (Princeton: Princeton University Press, 1895), 22–79.

169. Donna Haraway, *Modest_Witness@Second_Millennium.FemaleMan_Meets_OncoMouse: Feminism and Technoscience* (New York: Routledge, 1997), 23.

170. Cf. Peter Redfield, "A Less Modest Witness: Collective Advocacy and Motivated Truth in a Medical Humanitarian Movement on JSTOR," *American Ethnologist* 33, no. 1 (2006): 3–26.

experimentation.[171] Thinking that "one of the most unsatisfactory symptoms of the present day is the craving for incessant inspection in every function of life," Matthews rejected complaints over the limited extent of inspection made by MPs. Inspectors' presence at experiments, he claimed, was neither necessary nor possible.[172] Matthews's successor, Home Secretary Asquith, reacted similarly when MP Lockwood argued that the legislature intended personal inspection over experiments.[173]

Others proposed allowing representatives of the public to watch the experiments. Alpheus Morton, MP for Peterborough, expressed this idea during a debate in the House of Commons in 1895. Morton clarified that he did not wish to criticize the inspectors yet asserted that allowing interested people to witness experiments would reassure the public regarding whether there was any unnecessary cruelty involved. Home Secretary Asquith replied that he had no power to order public access to laboratories. The inspectors, who could visit the places under Parliamentary authority, were not authorized to allow entrance to anyone else. Asquith further argued that such an arrangement might interrupt experiments, which were often of a "very delicate nature."[174] Another unsuccessful proposal was presented by MP Lockwood later in 1898: "The laboratories should be open to a limited number of university graduates and medical men, who represent the public, when experiments are to be performed, in order that they may see that they are carried out in a humane manner, and under proper anæsthetics."[175]

Poore and Russell witnessed some operations even though the Act did not explicitly require them to do so. It had happened unintentionally during routine visits or occasionally following a command of the administration to watch an experiment. From 1899 to 1905, Thane or his deputy saw 100 operations.[176]

---

171. Memo on Inspection under the Cruelty to Animals Act, June 1902, HO 144/634/B37080.

172. 348 Parl. Deb (3rd ser.) (1890) 780.

173. 17 Parl. Deb (4th ser.) (1893) 345. In addition, the Home Office received repetitive requests to increase the number of inspectors or assistants, including a "numerously signed" petition in 1896. Letter on behalf of the Home Secretary to H. J. Reid, July 4, 1895, HO 156/9, 366; Memo, Inspection under the Cruelty to Animals Act, HO 144/634/B37080, [1902?].

174. 34 Parl. Deb. (4th ser.) (1895) 763. See also "Our Cause in Parliament: The Inspection of Vivisection Experiments," *Animal's Defender and Zoophilist* 15, no. 6 (July 1895): 205–6. The concept of a public representative in inspection was raised during the Royal Committee hearing. Samuel Haughton, a fellow at Trinity College in Ireland asserted that an inspection should be constituted from at least "three competent persons" and also that "the public must be represented on it." Royal Commission Report, 105.

175. 63 Parl. Deb. (4th ser.) (1898) 542.

176. Royal Commission on Vivisection, *Final report of the Royal Commission on Vivisection* (London: His Majesty Stationary Office, 1912), 8.

The inspectors were often intrigued by the scientific significance of the experiments, and, following the Act's requirements, they paid close attention to the workings of anesthesia. In a visit to the Physiological Department at University College in London, Poore observed Edward Schäfer engaged in a blood pressure experiment on a dog: "The animal was as unconscious as a dead animal."[177] In October 1895, Poore saw Vaughan Harley perform an enterostomy (creating an opening in the abdominal wall) on a dog, and noted that "the animal was necessarily deeply anesthetized."[178] Poore also saw J. S. R. Russell operating upon a dog at the Pathological Department at University College in London: "The dog was deeply anaesthetized and felt nothing and would be killed before recovery."[179]

Inspector Russell was instructed—by the home secretary or by Poore—to witness an experiment done by Gustav Mann at the Physiological Department at the University of Edinburgh. The procedure took an hour and a half, after which Russell concluded: "I do not think that the animal suffered appreciable pain and it seemed to me that on several occasions it nearly died from excessive doses of ether."[180] Russell's criticism of the use of anesthetics came up again when visiting the Royal College of Physicians of Edinburgh in June 1894. In one room he saw the physician Ralph Stockman bleeding a dog from a vein in the leg and then injecting the blood under the skin, probably as part of the investigation of iron-deficiency anemia.[181] Superintendent Diarmid Noel Paton, Dr. Miles, and two servants were also present. "The dog was much frightened and moaned and howled while under the influence of ether . . . The ether was required by the Act and was the cause of all the trouble by frightening the dog."[182] In this case it was the experimenter's adherence to the Act that made the animal suffer and yet, for the Home Secretary, experiments without anesthesia were more troubling. This is one of the very few occasions in which an inspector pointed to mental distress: the dog was frightened. Later that year the Home Office ordered Stockman to inform Russell in advance of the dates he planned to experiment without anesthesia.[183]

---

177. Poore's Report on 1894, 8.
178. Poore's Report on 1895, 29.
179. Poore's Report on 1894, 11.
180. Russell's Report on 1894, 2.
181. Ralph Stockman, "Observations on the Causes and Treatment of Chlorosis," *British Medical Journal* 2, no. 1824 (December 14, 1895): 1474. On the experimental investigations in the Laboratory of Royal College of Physicians in Edinburgh and its connection to clinical cases, see Steve Sturdy, "Knowing Cases: Biomedicine in Edinburgh, 1887–1920," *Social Studies of Science* 37, no. 5 (October 1, 2007): 659–89, 674.
182. Russell's Report on 1894, 3.
183. Secretary of State to Ralph Stockman, October 23, 1894, HO 156/9, 10.

The scientific knowledge produced by late nineteenth-century physiology not only had to be accurate but also ethical. The presence of inspectors in experiments as well as their inspection visits and oversight over laboratory records were needed to produce ostensibly cruelty-free experimental facts.[184] After overcoming their initial resistance to inspection, physiologists learned to use it as an ethical authorization. The presence of inspectors in laboratories, especially during experiments, aimed to reassure the British public that scientific knowledge was produced by ethically approved methods.

The civil servants at the Home Office were aware of the legitimatizing aspect of their work. Summarizing his laboratory inspections in 1895, Russell noted that some physiologists looked upon the Act "as conferring a measure of protection by shielding them from unfounded charges."[185] A civil servant in the Home Office, signed as W. P. B. (probably W. P. Byrne), demonstrated a similar understanding of the role of inspection when he claimed that it was an "undoubted fact that it is not the inspection which makes experimenters carry out their operation with due humanity." The only reason to employ a more frequent inspection (which he resisted) was to "allay the public anxiety."[186]

Shortly after his resignation in October 1899, inspector Poore delivered the Harveian Oration at a meeting of the Royal College of Physicians of London. His speech revealed an unsurprising alliance with the experimenters and his discontent with the antivivisection critique: "There be those who apparently hold the view that a guinea-pig is of more value than many babies ... With such as these it is useless to argue. But seeing that many honored members of our profession have themselves been vivisected by the envenomed tongues and sharp pens of a few noisy people, it may be well to point out that no conviction for cruelty or breach of law has ever obtained."[187]

Poore cited the absence of legal convictions as proof of the experimenters' proper conduct—a perplexing statement in that he was the chief person whose reports could have provided the basis for a prosecution. Critics of the Act used Poore's speech as evidence of Home Office biases in enforcing the Act. Stephen

---

184. This dynamic resembles the role of state veterinarians in regulating American meat production, described by the historian Susan Jones as an attempt "to reconcile the exploitation of animals for food, work and companionship with America's need to feel morally comfortable with those uses." Susan D. Jones, *Valuing Animals: Veterinarians and Their Patients in Modern America* (Baltimore: Johns Hopkins University Press, 2002), 9.

185. Russell to Poore, March 5, 1896, HO 144/383/B19846A.

186. Memo on Inspection under the Cruelty to Animals Act, June 1902, HO 144/634/B37080.

187. George Vivian Poore, "The Harveian Oration," *British Medical Journal* 2, no. 2025 (October 21, 1899): 1106.

Coleridge from the Victoria Street Society published a response in the *Daily Chronicle*, arguing that Poore had "done the cause of antivivisection an immense service, for, as he is the person appointed by the government to inspect the laboratories of the vivisectors, the public can judge for themselves the quality of the impartiality he brings to the task!"[188] Coleridge also rightly observed in 1900 that the inspectors' reports "had for many years past chronicled no single instance of appreciable pain having been inflicted on any animal, in any laboratory, by any vivisector."[189] However, Poore's statement was not representative of all inspectors, and very different from those made, for example, by Thornley Stoker, inspector for Ireland for three decades. Stoker testified before the second Royal Commission that he experienced "a growing sense of humanity, if I may call it so. I found myself for many years past having a growing sense of appreciation of the suffering of the lower animals."[190]

Despite the inspectors' general support of the physiological enterprise, it would be wrong to describe the inspectors merely as agents of the scientific establishment, embodying the interests of vivisectors inside the state apparatus. The visitation reports, notes, and letters of the inspectors show they were figures of both scientific and bureaucratic worlds who became agents of coproduction. When the inspectors entered late nineteenth-century physiology laboratories, they guided the experimenters on how to produce knowledge lawfully and in line with contemporary humanitarian sensibilities to animal pain—that is, with the ostensibly minimal infliction of pain necessary to achieve the objects of the experiments. They formally and informally advised physiologists on the Act's interpretation, and they enforced its requirements for vivisection practitioners to obtain licenses and certificates for their work. The inspectors also exercised judgments about animal pain and the proper conditions for an experimental space, concurrently compelling the physiologists they visited to engage with the same questions. They facilitated the incorporation of a moral-legal perspective on animals into the working of laboratories. Their work shows how, under the Vivisection Act, scientific practice and legal standards coproduced knowledge and normative order.

---

188. Stephen Coleridge, "The Vivisection Inspector on Ourselves," *Animal's Defender and Zoophilist* 19, no. 7 (1899): 142.

189. Stephen Coleridge, An Open Letter to the Right Hon. Secretary of State, December 1900, HO 144/606/B31612.

190. Appendix to First Report, 34.

5

LIBEL, SLANDER, AND VIVISECTION

Louise (Lizzy) Lind-af-Hageby and Liese Schartau were probably the first undercover antivivisection agents in British history. In 1902, Lind-af-Hageby, an honorary secretary of the Swedish Anti-Vivisection Society, and her friend Schartau went to study physiology in England, aiming "first, to investigate the *modus operandi* of experiments on animals, and then to study deeply the principles and theories which underlie modern physiology."[1] The curriculum at the London School of Medicine for Women—the first medical school in Britain to train women as doctors—did not include vivisection. Nevertheless, during the autumn of 1902 and the winter of 1903, Lind-af-Hageby and Schartau were offered the opportunity to observe a series of physiological demonstrations at King's College, Imperial Institute, and University College.[2] The two would have probably rejected their description as undercover agents at the medical school. Lind-af-Hageby, the more dominant of the two, insisted that neither she nor her partner hid their beliefs or political associations: "We have given our full names and addresses and paid our fees. Whenever medical students have spoken to us about vivisection, we have made no secret of our opinions. If anybody had cared to enquire, our connection with the Scandinavian Anti-Vivisection movement could easily have been found out."[3] But nobody cared

---

1. Lizzy Lind-af-Hageby and Leisa Katherina Schartau, *The Shambles of Science: Extracts from the Diary of Two Students of Physiology* (London: E. Bell, 1903), vii.

2. *Oxford Dictionary of National Biography*, s.v. "Lind-af-Hageby, (Emilie Augusta) Louise (1878–1963)," by M. A. Elston, http://www.oxforddnb.com/view/article/40998; Lansbury, *The Old Brown Dog*, 9. For more about Lind-af-Hageby, see Hilda Kean, *Animal Rights: Political and Social Change in Britain since 1800* (London: Reaktion Books, 1998), 140; Marc Bekoff and Carron A. Meaney, ed., *Encyclopedia of Animal Rights and Animal Welfare* (New York: Routledge, 2013), 234.

The most comprehensive text about Lind-af-Hageby is in Evalyn Westacott, "A New Leader—Miss Emilie Augusta Louise Lind-Af-Hageby," in *A Century of Vivisection and Anti-Vivisection: A Study of Their Effect upon Science, Medicine and Human Life during the Past Hundred Years* (Ashingdon, Essex: C.W. Daniel Co., 1949), 189–96.

3. Lind-af-Hageby, *The Shambles*, xi.

to inquire while the Swedish students "vividly and dramatically documented scenes of vivisection."[4]

Lind-af-Hageby and Schartau's documentation resulted in the publication in 1903 of the sensational *The Shambles of Science: Extracts from the Diary of Two Students of Physiology*, and in a consequent libel case by the physiologist William Bayliss, known as the "brown dog affair." As this chapter shows, formally the Home Office was not a party to the libel dispute, yet it closely followed as the event unfolded. The main concern of the Home Office was that the *Shambles* was not only an indictment of physiological research, but also an implicit critique of the Home Office's implementation of the Act. All the experiments described in the book were performed in registered places, and their perpetrators held licenses duly signed by the Home Office. Violation of the law on the scale described by the book left the Home Office with much to explain. It therefore ordered inspector Thane to inquire into the book, who took pains to debunk each of the book's claims of animal suffering. The libel trial triggered by the *Shambles* was one of the factors leading to the establishment in 1906 of the second Royal Commission on Vivisection to examine the Act, where Lind-af-Hageby was considered a "most important witness," and Thane's analysis of the *Shambles* served as a central document.[5] The commission's report displayed the developments in British scientists' understanding of pain management, as well as their acceptance of regulation. Its recommendations, however, did not include legislative revisions and focused on modifications in the Act's administration. The chapter ends with a libel action initiated by Lind-af-Hageby at the High Court, which shows not only the centrality of the Act in setting the terms in which vivisection was debated in civil contexts, but also how the interpretation of animal pain expanded to leaflets' representations and stuffed dogs.

Bayliss v. Coleridge

Publishing their observations was not a part of Lind-af-Hageby and Schartau's initial plan, yet in April 1903, with their notebooks filled with the accounts of

---

4. Lansbury, *The Old Brown Dog*, 10.

5. Elston, "Women and Anti-Vivisection," 285; Royal Commission on Vivisection, "Final Report of the Royal Commission on Vivisection" (London: Printed by Wyman & Sons for H.M. Stationery off, 1912), 16 (hereinafter: "Final Report").

about fifty experiments, they began drafting a manuscript.[6] They reached out to Ernest Bell, a publisher and an animal welfare campaigner.[7] Bell told Stephen Coleridge, honorary secretary of the National Anti-Vivisection Society, about the manuscript, and the latter asked Lind-af-Hageby to see it. She arrived at his home one morning and read him "one or two" chapters.[8]

Of the various experiments described in Lind-af-Hageby and Schartau's notes, Coleridge was most alarmed by a demonstration performed by the physiologist William Bayliss on February 2, 1903. According to Lind-af-Hageby and Schartau, the subject of the experiment, a brown dog, bore the marks of a previous vivisection. Moreover, they inferred that the dog was not anesthetized since it moved on the operating table, and because when they approached it at the end of the demonstration, they did not see any tubes attached to his body, nor could they detect a smell of ether or chloroform.[9]

The demonstration was particularly unsettling for Coleridge since it was performed, according to the manuscript, in an atmosphere of lightheartedness, greeted by the student crowd with bursts of laughter. In order to publicize the event, Coleridge asked Lind-af-Hageby for a written statement. At first, she declined, claiming she did not intend to attack individuals "but to attack the system." She also did not want to publish anything before the whole book was out. But eventually she was persuaded, being told that "in this controversy it is of very little use to make vague statements but that it is far better to put names and addresses to one's charges."[10]

Coleridge made the statement public at the annual meeting of the Society at St. James's Hall on May 1, 1903. He was aware of the possible consequences of his speech, as he reassured his audience, "I have not allowed myself to make this statement public until I had ascertained from both these persons that they were ready, if necessary, to substantiate every syllable that they had stated upon oath in the witness-box."[11] He moved on to read from the statement:

---

6. Lind-af-Hageby, *The Shambles*, xi; extracts from the *Times*, November 12, 1903, HO 144/589/B7733.

7. In the High Court of Justice Kings Bench Division, Before: Mr Justice Bucknill and a Special Jury Lind-af-Hageby v Astor and others, Third Day, April 3, 1913, WA, GC/89/1, 31.

8. Lind-af-Hageby v Astor, Third Day, 32. Caitlin Harvey analyzed the manuscript as a peculiar medium, a diary that transgressed contemporary gender scripts of female antivivisectionists as sentimental and uninformed. Caitlin Harvey, "Science and Sensibility: Louise Lind-Af-Hageby's Diary as Female Testimony, Scientific Publication, and Antivivisectionist Tool, 1890–1918," *Journal of Women's History* 30, no. 1 (2018): 80–106.

9. Extracts from the *Times*, November 18, 1903, HO 144/589/B7733.

10. Lind-af-Hageby v Astor, Third Day, 32, 33.

11. Extracts from the *Times*, November 12, 1903, HO 144/589/B7733.

A big brown dog of terrier type was brought into the lecture-room stretched on its back on the operation board ... it was muzzled so tightly that it was now deprived of every power to give audible expression to its pain. In the skin of the abdomen there were several scarcely-healed scars and wounds; in one of them, that seemed to be rather fresh, there were left a pair of clamping forceps. It was evidently not the first time that this dog had had to serve science. The internal organs of the abdomen had surely had their turns of operations in previous experiments. The neck was opened widely for the stimulation by electricity of a certain gland. The dog struggled forcibly during the whole experiment and seemed to suffer extremely during the stimulation. No anaesthetic had been administrated in my presence.[12]

The speech was published in the *Morning Post* and the *Times*.[13] The *Daily News* urged action, "those concerned must either refute it or admit its accuracy."[14] MP Frederick Banbury asked to know "what was the nature and amount of the anaesthetic given to the brown dog," as well as how long before its entrance to the theater the animal was anesthetized, who operated upon the dog before Bayliss, and under what kind of a certificate.[15] MP Lockwood inquired whether the home secretary was aware that the brown dog had undergone three operations, and whether there was "any official report showing that the animal was placed under anaesthetic and had suffered no ill effects from the operation."[16] Most queries were forwarded to inspector Thane to draft the reply. The official reply was based on letters and statements from Starling and Bayliss, and contended that the animal was under anesthetics during each of the operations and was killed while anesthetized.[17]

On July 10, 1903, Bell published Lind-af-Hageby and Schartau's journals under the title *The Shambles of Science: Extracts from the Diary of Two Students of Physiology*. The authors omitted the names of lecturers and demonstrations, "as this is not meant to be a personal attack, but an indictment against the system."[18]

---

12. Extracts from the *Times*, November 12, 1903.

13. Extract from the *Morning Post*, May 2, 1903, HO 144/589/B7733; Extracts from the *Times*, May 4, 1903, HO 144/589/B7733.

14. Extract from the *Daily Mail*, May 2, 1903, HO 144/589/B7733.

15. Home Office Memo, May 18, 1903, HO 144/589/B7733; 122 Parl. Deb. (4th ser.) (1903) 1200.

16. Home Office Memo, May 12, 1903, HO 144/589/B7733. See also the question of Mr. Weir on June 11, 1903 about the number of dogs used without anesthesia at University College London by Bayliss during 1902, in 123 Parl. Deb. (4th ser.) (1903) 641.

17. Statement signed by Bayliss, May 8, 1903, HO 144/589/B7733; Straling to Thane, May 25, 1903, HO 144/589/B7733.

18. Lind-af-Hageby, *The Shambles*, xii.

FIGURE 7 / Sir William Bayliss and the reconstruction of the brown dog experiment, c. 1900. Wellcome Collection, L0029652.

Shambles was an immediate success, and Lind-af-Hageby reported it received "more than two hundred notices and reviews in daily papers and other journals" in the first four months after its publication.[19] A reprint of the book was published as soon as August 1903.

Facing the mounting outcry, Bayliss filed a lawsuit against Coleridge, "to recover damages for libel and slander and for an injunction to restrain further publication." The *Bayliss v. Coleridge* trial commenced on November 11 and lasted four days. Bayliss's main argument was that the brown dog was adequately anesthetized. Relying on the Act, the plaintiff stressed that there was "no cutting operation that was licensed by the Secretary of State unless it was done under anaesthetics so that the animal did not suffer pain." Bayliss and his partner in the experiments, Ernest Starling, confirmed that the dog moved during the operation, but argued that the movement was produced by "a previous attack of distemper."[20] Female students were also called to testify in support of Bayliss,

---

19. Précis of Statement of Evidence to be given by Miss L. Lind af Hageby, HO 114/4, 3.

20. Extracts from the *Times*, November 12, 1903, HO 144/589/B7733.

a move one commentator interpreted to be a deliberate strategy "to defuse the sexual antagonism" implicit in the accusations made by Lind-af-Hageby and Schartau.[21]

Bayliss won the case and was awarded £2,000 in damages—"a large sum, almost a sensational amount in the circumstances," according to the *Edinburgh Evening News*.[22] The *British Medical Journal* celebrated the victory over Coleridge, the "protomartyr of antivivisection" who was deceived by "two Swedish ladies, whom he must have known to be biased witnesses."[23] The *Scotsman* rejoiced: "fanatics must sometimes pay the penalties of fanaticism," and the *Glasgow Daily Record and Mail* warmly congratulated the "exemplary verdict" against Coleridge and the foreigners whose "real purpose" was "to spy out the land."[24]

The *Westminster Gazette* commented on "the inadvisability of ladies who adopt the medical profession allowing themselves to be carried away by too much sentiment, and ignoring the fact that a serious responsibility attached to the evidence of those who are in any way qualified to speak as experts."[25] The newspaper also denounced "the jocularly of the students" about which Lind-af-Hageby and Schartau reported. The verdict gained overwhelming support also from the *Times*, the *Standard*, the *Morning Post*, the *Daily Telegraph*, and the *Daily Chronicle*.[26] Conversely, the *Daily News* accused the jury of ignoring the uncontroversial fact that the dog had been vivisected three times.[27]

Following the decision in *Bayliss v. Coleridge* and in response to Bayliss's solicitor's threat of legal action, Bell undertook to withdraw *Shambles* from circulation.[28] He furnished Bayliss's solicitors with the following letter:

> I . . . hereby acknowledge that I have given instructions for the withdrawal from circulation of all copies of such book, and hereby undertake that no further copies of such book shall be printed or published by me; that the circulation of such book shall cease; and that all copies in stock and withdrawn from circulation shall be handed over to Messrs.

---

21. Mary Ann Elston, "Women and Anti-Vivisection in Victorian England, 1870–1900," in *Vivisection in Historical Perspective*, ed. Nicolaas A Rupke (London; New York: Routledge, 1990), 285.

22. Extracts from the *Edinburgh Evening News*, November 19, 1903, HO 144/589/B7733.

23. "The Protomartyr of Antivivisection," *British Medical Journal* 2, no. 2239 (1903): 1415.

24. Extracts from the *Scotsman*, November 19, 1903, HO 144/589/B7733; extracts from the *Glasgow Daily Record and Mail*, November 19, 1903, HO 144/589/B7733.

25. Extracts from the *Westminster Gazette*, November 19, 1903, HO 144/589/B7733.

26. Paper clips, HO 144/589/B7733.

27. *Daily News* November 19, 1903, HO 144/589/B7733.

28. Précis of Statement of Evidence to be given by Miss L. Lind af Hageby, HO 114/4, 3.

Hempson, Dr. Bayliss solicitors, and I hereby express to Dr. Bayliss my sincere regret for having printed and published the book in question.[29]

The *Shambles* authors opposed the move, and with the help of friends to the antivivisection cause they raised enough funds for a fourth edition of *Shambles*. The July 1904 edition omitted the controversial chapter "Fun," which described Bayliss's experiment on the brown dog.[30]

The brown dog affair outlived the Bayliss case in more than one way. Following the trial antivivisectionists erected a memorial to the brown dog in Battersea, with the plaque:

> In memory of the brown terrier dog done to death in the laboratories of University College in February 1903, after having endured vivisection extending over more than two months and having been handed over from one vivisector to another till death came to his release. Also in memory of the 232 dogs vivisected at the same place during the year 1902. Men and women of England, how long shall these things be?[31]

Protests followed the erection of the statue, involving clashes between medical students, suffragists, and working-class men.[32] While the dispute moved to the streets, the legal and administrative involvement of the Home Office in the *Shambles* quietly deepened. The inspectorate was mobilized to check, if not to refute, the testimonies recorded by the *Shambles*. The Home Office's rereading of the testimonies was a telling exercise in the interpretation of animal pain.

Behind the Scenes of the Brown Dog Affair

A deputation of antivivisectionists called the Home Office's attention to the book soon after its publication. The home secretary responded that "the statements that licensed practitioners were indifferent to the suffering of the animals, or that any of them enjoyed the spectacle of suffering, was a gross and

---

29. The letter, dated November 25, 1903, was read by Thane during his testimony at the second Royal Commission. Royal Commission on Vivisection, "Appendix to First Report of the Commissioners: Minutes of Evidence, October to December, 1906" (London: printed for H.M.S.O. by Wyman & Sons, 1907), 49.

30. "Final Report," 16; Précis of Statement of Evidence to be given by Miss L. Lind af Hageby, HO 114/4, 4.

31. Linda Kalof, *Looking at Animals in Human History* (Reaktion Books, 2007), 140.

32. The class and gender dimensions of the demonstrations are examined in Lansbury, *The Old Brown Dog*.

grotesque libel on a humane, self-sacrificing, and high-minded profession."[33] But as in previous occasions, the Home Office publicly defended physiologists while at the same time initiating an internal inquiry into the accusations against them. A brief correspondence between inspector Thane and the physiologist Augustus Désiré Waller regarding the latter's experiments at the University of London then took place.[34] Thane informed the Home Office that he did not think it necessary to take any further steps regarding Waller's experiments.[35]

The *Bayliss Case* erupted a month later and forced the Home Office to study the book more closely and to prepare a detailed response. William Byrne, then a principal clerk at the Home Office, was notified that Home Secretary Aretas Akers-Douglas "will be glad if a Report (which may have to go to the King) can be drawn up of the actual facts of each of the experiments which the 'Shambles of Science' purports to describe. It should, if possible, be in such a form as to furnish answers to the various statements or insinuations."[36]

The interrogation mission was assigned to inspector Thane, who devoted a few weeks to examining the episodes described in the book.[37] Other officials at the Home Office worked in parallel to inquire and collect official responses to the *Shambles*, including the reinvestigation of Waller's experiments. On November 26, 1903, legal assistant undersecretary Cunyngham asked A. W. Rücker, principal of the University of London, if the latter was aware of the *Shambles*'s allegations. He requested Rücker to inform Home Secretary Akers-Douglas whether he "wish to make any observations on the statements made in the book."[38] Rücker reassured Cunynghame that early in October, before the trial began, the director of the Physiological Laboratory had carefully examined the allegations related to the university, and found them "in every essential particular, incorrect."[39] He added that the university decided to take no action against the *Shambles'* authors at that time in view of the pending action for libel in *Bayliss v. Coleridge.*

---

33. Memo, July 1903, HO 144/606/B31612. A copy of the *Shambles* was received at the Home Office on July 17, 1903, in HO 144/606/B31612.

34. Waller to Thane, August 8, 1903, HO 144/606/B31612.

35. Memo, August 8, 1903, HO 144/606/B31612.

36. I.G. [?] to Byrne, November 2 [?], 1903, HO 144/606/B31612.

37. Royal Commission on Vivisection, "Appendix to First Report of the Commissioners: Minutes of Evidence, October to December, 1906" (London: printed for H.M.S.O. by Wyman & Sons, 1907), 47.

38. Cunynghame to Sir A. W. Rücker, November 26, 1903, Experiments in the Physiological Laboratory of the University, HO 114/2.

39. Rücker to the Under Secretary of State, December 3, 1903, Experiments in the Physiological Laboratory of the University, HO 114/2.

To demonstrate the university's commitment to the legal process, Rücker called the Home Office's attention to a recent *Lancet* publication by Waller. It was the text of a lecture that Waller had delivered a month and a half earlier in the Physiological Laboratory at the University of London, entitled "The Administration of Chloroform to Man and to the Higher Animals."[40] The delay in the publication, explained Rücker, "was due to the desire of the University authorities not to take any improper step while the libel action was pending or in course of trial." Rücker clarified that as the court had discredited the *Shambles* allegations, the university did not plan to take any further actions. He also informed the Home Office that when the laboratory was first established and on several other occasions, he had informed Waller of his "personal wish that no such experiments should be used for the purposes of demonstration except those which were necessary for the instruction of senior students," and that Waller had concurred.[41]

Thane submitted his "Observations on 'The Shambles of Science'" on December 5, 1903.[42] The document, designated "confidential," opened with a general validation of Lind-af-Hageby and Schartau's accounts: "Although there are many exaggerations, and sometimes false statements, arising either from misapprehension or ignorance, the accounts given of the proceedings themselves are generally fairly correct." The problem, according to Thane, was the delivery and the overtly emotional interpretation of the events: "What is not right or just, however, is the tone in which the work is written, the suggestion of dishonesty and dishonorable conduct, and of brutality on the part of the lecturers and their assistants, and the constant interpolation of extraneous matter, either culled from other sources, or expressing the feelings and emotions of the writers with regard to experiments in general, all narrated as if they were actual occurrences in the events dealt with."[43] Thane went on to dissect the *Shambles* chapter by chapter, often siding with the physiologists' versions of the events. He based his findings on interviews with experiments' participants, laboratory records, his personal experience, and on a textual analysis of the book.

The *Observations* dismissed Lind-af-Hageby and Schartau's impressions on the grounds of overt emotionality or ignorance. For example, the chapter "Painless Experiments" in *Shambles* described an experiment on the nervous

---

40. Augustus D. Waller, "A Lecture on The Administration of Chloroform to Man and to the Higher Animals, Reprinted from the Lancet," November 28, 1903, HO 114/2.

41. Rücker to the Undersecretary of State, December 3, 1903, HO 114/2.

42. G. D. Thane, "Observations on 'The Shambles of Science,'" HO 144/606/B31612.

43. Thane, *Observations*, 1.

system and body temperature undertaken at Imperial Institute on December 2, 1902. As part of the demonstration, a rabbit "was put in a freezing-machine where a big piece of ice had been put previously." According to the *Shambles* account, the rabbit was "left too long in the box, and when it was taken out after fifty-five minutes, it was found to be 'beyond the stage for observation.'" The rabbit was, according to the authors, "quite conscious but frozen stiff, like a piece of wood. With all signs of terror the animal springs back, trying to get away" but "half-paralyzed by the cold" it collapsed.[44] In *Observations*, however, Thane contended that the rabbit was anesthetized before it was put into the cool chamber, emphasizing he had seen the specified cooling device, and found it to be "an ordinary refrigerator." He dismissed the description of the frozen rabbit as "absolutely untrue," since "an animal frozen stiff like a piece of wood cannot spring back with all the signs of terror; nor can an animal that is frozen still be quite conscious."[45]

However, Thane accepted the descriptions in *Shambles* as reliable evidence when it suited his argument. The book's chapter "Blood-Clotting" opened with a favorable account, exceptional in antivivisection literature, of administration of anesthesia to a gray rabbit during a blood clotting experiment at the University of London. It describes how "the anaesthetist looks most careful, as if he were attending the precious life of some human patient, and to many this would be a perfect and satisfying demonstration of the extreme care, nay love, with which the animals are treated in the laboratories. These moments have fixed themselves in our memory with a persistency that can only be explained by the fact that it is the only time that we have witnessed such a tender ministry." However, later in the experiment, the rabbit "begins to struggle as much as it can with its tied feet; it is then thrown into tetanic convulsions for about a minute, the eyes become bloodshot and start from their sockets." The animal seemed to be dead, only to show some signs of struggle later.[46]

The "most careful" administrator of anesthesia was revealed by Thane to be Waller, the director of the laboratory. Thane was satisfied that the anesthesia was adequate, reasoning that the animal must have been still to insert the cannula into the vein in its neck. Since according to the *Shambles* the rabbit's head was not sufficiently fixed to prevent it from moving it, "the stillness must therefore have been due to the anaesthetic." The rabbit's struggle and tetanic convulsions were "gross exaggerations." Drawing from a study by the physiologist William

44. Lind-af-Hageby, *The Shambles*, 13, 14.
45. Thane, *Observations*, 2.
46. Lind-af-Hageby, *The Shambles*, 31.

Halliburton, Thane concluded that what the authors characterized as struggles were probably "the stretching movements and the lost of breaths," tetanic spasms that did not indicate a consciousness or sensation. He further dismissed the significance of evidence of animals moving while being dissected, reasoning, "the muscles do not die immediately with the individual."[47]

Thane knew some of the animals firsthand. The "Quiet Cat" chapter in *Shambles* opened with the authors mentioning once seeing a marmot, "the spinal cord of which had previously been divided, bite a vivisector."[48] The chapter focused on an experiment with blood pressure on a black and white cat done at University College on February 9, 1903. Thane, however, was fixated on the marmot's comment. "I know this marmot well," wrote Thane, "I have often seen it." Thane contended that it was never experimented upon save for having its temperature taken. Noting that it was the only marmot that had been shown to physiologists in Britain, Thane regretted the death of the marmot. He examined its body and claimed that contrary to the claims in *Shambles*, the marmot's spinal cord had not been divided, and the paralysis described in *Shambles* was hibernation. This phenomenon "resembles the state when the spinal cord has been divided; but nothing of the sort had been done here." This was "a very good example of the way in which the most innocent proceeding is misrepresented."[49]

In the chapter "Scarcely Any Anaesthesia," the *Shambles* authors described a brown terrier undergoing a brain stimulation experiment: "The left eye is squeezed together against the table; when we bend down to look closely at the right one there is a look of the utmost agony in it."[50] Thane, however, questioned the existence of an agonized look: "The look of utmost agony in the eye is the imagination of the writer, as emotions are not expressed in the eye itself, but in the surrounding parts; the look of agony in the eye is a figure of speech."[51] Similarly to his dismissal of the authors' interpretation of the rabbit's spasms, Thane cast doubt on their empathetic process. The sights they were describing were perhaps accurate, but their interpretation of the indications of pain was flawed, exaggerated, and unskilled.

Referring to the description that the terrier's leg was "still and motionless, as if it were part of a dead dog," Thane said it "reads like a paragraph from an inspector's report; only it would probably not be said there that the limb

47. Thane, *Observations*, 2, 3.
48. Lind-af-Hageby, *The Shambles*, 37.
49. Thane, *Observations*, 3.
50. Lind-af-Hageby, *The Shambles*, 50.
51. Thane, *Observations*, 4.

was stiff; under the circumstances, the limb would be flaccid."[52] With these comments—excluding from it "figures of speech" and replacing the word "stiff" with "flaccid"—Thane demonstrated how the text could have been edited to turn it into an inspection-like, allegedly objective, depiction of the brown terrier under experimentation.

In the closing remarks of *Observations* Thane identified the two main concerns of the *Shambles* as an inadequate use of anesthetics and "the tone of the proceedings—the spirit in which they are conducted." He contested the issue of anesthetics misuse systematically throughout his commentary, but as an additional support he emphasized that eight out of the thirteen demonstrations described in *Shambles* took place in Waller's laboratory, whom Thane judged to be, as "one may say, an expert, probably the first in the country, on the physiology of anaesthetics." For inspector Thane, Waller's scientific expertise had a priori guaranteed that the experiments were done with proper anesthetics as required by law. Drawing on his own experience, Thane also rejected the accusation of the entertaining aspect of the vivisection demonstrations. He claimed that he was present at one of the events described in the book, that he was "well acquainted with some of the students," and that "this is not at all their character. The lectures here referred to were all on 'Advanced Physiology,' and the students attending them were all seriously and earnestly seeking information and instruction." Furthermore, Thane used the court decision in *Bayliss v. Coleridge* to shed doubt on Lind-af-Hageby and Schartau's reporting: "The matter came up at the recent trial, when one of the authors was cross-examined as to 'Fun,' and the absence of foundation for the allegation was clearly shown."[53]

The Second Royal Commission

In early March 1906, Amelius Mark Lockwood, a Conservative MP for Epping, addressed the Home Secretary at the House of Commons with the request that dogs, "which had a higher nervous organization than other animals," would be prohibited from vivisection. His comment that he "would say nothing about the guinea-pigs and mice; but he did plead for the dogs,"[54] was greeted with cheers.[55] Lockwood added that if the Home Secretary "could not see his way to exclude

---

52. Thane, *Observations*, 7.
53. Thane, *Observations*, 9.
54. 153 Parl. Deb. (4th ser.) (1906) 164.
55. Extracts from the *Times*, March 6, 1906, HO 45/10521/138422.

dogs from the operation of the Act," he requested, "in view of all that had taken place since the original Royal Commission sat, and in view of all the evidence adducible in this country and in various parts of the United States, to appoint a Committee, or an impartial Commission which should consider the evidence on both sides" and make recommendations to the House.[56] A week later another MP urged the establishment of a new commission, and Home Secretary Herbert Gladstone concurred on the grounds of changes in medical research: "The present law was based on the Report of the Royal Commission of 1876. Since then there have been great changes in medicine and the methods of scientific research. I assent to the view that it is desirable that an inquiry should be held, and I will consider what form the inquiry should take."[57]

On September 17, 1906, Gladstone appointed the Royal Commission on Vivisection "to inquire into and report upon the practice of subjecting live animals to experiments, whether by vivisection or otherwise; and also to inquire into the law relating to that practice, and its administration; and to report whether any, and if so what, changes are desirable."[58] Gladstone appointed ten commissioners, of which four, "including two distinguished medical men, were suggested to him by anti-vivisection societies." He refused, however, to provide a curious MP with the names of the commissioners who had been recommended by the Anti-Vivisection Society.[59]

The first chairman of the committee was William Court Gully, the first Viscount Selby, a lawyer and a Liberal who was the speaker of the House of Commons from 1895 to 1905. The original committee members were MP Lockwood; William Selby Church, a physician and a former president of the Royal Medical and Chirurgical Society; William Job Collins, a surgeon and Liberal MP for West St Pancras; John McFadyean, president of the Royal College of Veterinary Surgeons; Mackenzie Dalzell Chalmers, a former judge and the permanent undersecretary at the Home Office; Abel John Ram, a barrister; Walter Holbrook Gaskell, a physiologist; James Tomkinson, Liberal MP for Crewe; and George Wilson, a doctor of medicine. Clive Bigham, C.M.G, was the commission's secretary.[60] In 1909, Abel John Ram replaced the deceased Gully as chair;

---

56. 153 Parl. Deb. (4th ser.) (1906) 164.

57. 153 Parl. Deb. (4th ser.) (1906) 1099.

58. "Final Report," ii.

59. 163 Parl. Deb. (4th ser.) (1906) 419.

60. A dispute involving Gaskell was described by Lind-af-Hageby. In the High Court of Justice Kings Bench Division, Before: Mr Justice Bucknill and a Special Jury Lind-af-Hageby v Astor and others, Third Day, April 3, 1913, WA, GC/89/1, 7.

commissioner Tomkinson passed away but was not replaced; and Wilson was sick for a substantial period.[61]

The commission operated for six years during which it held more than seventy meetings, processed numerous papers, and examined representatives of governmental departments, universities, private practice, and antivivisection and humane societies.[62] In a controversial move, the commission's meetings were closed to the public and the press was not allowed, yet it promised to supply copies of the evidence "from time to time to certain representative bodies interested in the inquiry."[63]

Most of the scientists' evidence submitted to the second Royal Commission was dedicated to conveying the effectiveness of vivisection. Thus, for example, the president of the Royal College of Surgeons of England submitted a statement surveying the "advances in surgical knowledge which have been made, or, in part made, since 1876, by experiments on animals."[64] The commission noted that, in contrast to the situation faced by the commission of 1875, "there can be no doubt that the great preponderance of medical and scientific authority is against the opponents of vivisection."[65]

In addition to being united in their belief in the necessity of vivisection, most scientists testifying before the commission viewed the Act favorably. For example, the University of Manchester praised its cooperative relations with the inspectors. It wished to leave the state of affairs as it was, claiming that "adequate care is taken to further those objects which the Act was framed to accomplish and that the present legislative provision for controlling this type of investigation is sufficient."[66]

Nonetheless, despite the overall positive tone, vivisectors would have been happy to see some changes in the Act, in particular with regard to the limitations on demonstrations of experiments in classes. The Home Office presented to the commission requests by scientists to ease the requirement for a certificate

---

61. George Wilson, "Reservation Memorandum," "Final Report," 74.

62. Governmental representatives were not only from the Home Office. For example, the Local Government Board provided evidence regarding its experiments. Précis of Evidence to be given by Mr. W.H. Power, C.B., FRS, Medical Officer of the Local Government Board, n.d., HO 114/4.

63. 163 Parl. Deb. (4th ser.) (1906) 419. This was not unheard of. As Gladstone explained, "in several recent cases the evidence has been taken in private; I may instance the Motor Car Commission, the War Commission of 1902, and the Royal Commission on Food Supply in Time of War." 163 Parl. Deb. (4th ser.) (1906) 1108.

64. Statement Prepared for the Royal Commission on Vivisection by the President of the Royal College of Surgeons of England, n.d., HO 114/4.

65. "Final Report," 47.

66. Précis of Evidence Before the Royal Commission, n.d., HO 114/4.

C, which was allowed only when the animal was under complete anesthesia. This was particularly problematic in the case of inoculation (see chapter 3). However, there was not enough evidence to convince the commission that the Act's requirements "occasion any serious impediments to the general progress of science" and therefore, it did not recommend any amendments regarding experiments during lectures.[67]

The animal protection societies pronounced the strongest calls for amending the Act and the fiercest critique of its implementation. The establishment of the commission was their first opportunity in thirty years of accumulating data to lay out their reservations before an official tribunal. They were, at first, enthusiastic; Coleridge provided the commission with a list of issues that he considered essential. Prominent among them was the identification of pain in curarized animals: "To inquire into the nature of the drug curare and to advise whether in view of its alleged effects upon the motor nerves independently of the sensory nerves and the alleged consequent difficulty for any inspector to recognize the presence of agony in an animal subjected to its influence, the drug should be entirely prohibited."[68] He also submitted a list of suggested witnesses, including himself, from which the commission asked him to select "two witnesses to speak to the medical or physiological, and two to the ethical side of the question."[69]

The antivivisection groups soon found out that they had to curtail their hopes. The procedures designed by the commission signaled that it would not provide the public stage that these societies desired. The British Union for the Abolition of Vivisection had refused to take part in the proceedings unless a medical representative of the antivivisectionists was appointed to the commission.[70] The National Anti-Vivisection Society, led by Coleridge, joined the call.[71] Furthermore, the society protested the commission's decision to close its meetings to the public, and not to allow the society's counsel to take part in the hearings. Coleridge contended that "fullest publicity, and the examination and cross-examination of the witnesses by Counsel are both essential for the proper elucidation of the truth about vivisection as practiced in this country."[72]

---

67. "Final Report," 54.
68. Coleridge to Soares, May 25, 1906, HO 114/2.
69. Coleridge to the Secretary of the Commission, October 20, 1906, HO 113/4; Secretary of the Commission to Coleridge, October 25, 1906, HO 113/4.
70. John Stuart Verschoyle to Herbert Gladstone, October 12, 1906, HO 114.2; Unsigned letter (Home Office paper) to Bigham, November 7, 1906, HO 114/2.
71. Coleridge to Wilson, November 2, 1906, HO 114/2.
72. Coleridge to Captain C. Bigham, October 29, 1905, HO 114/3.

FIGURE 8 / "Anxiously awaiting their fate," undated newspaper clipping that refers to the decision made by the second Royal Commission on Vivisection to hold private meetings. The sign reads: "Meetings in secret, facts too terrible to make public." TNA, HO 114/1.

Consequently, Coleridge announced that the society "will not in any way participate in the proceedings of the Royal Commission unless the press are admitted and Counsel permitted to appear on its behalf."[73] The commission

73. Coleridge to Captain C. Bigham, October 29, 1905, HO 114/3.

rejected the society's ultimatum.[74] Coleridge complained about the issue to commissioner Wilson, whom he supported for the position: "I naturally looked to you to struggle valiantly for such rules of procedure," since "in my opinion it is quite impossible for you or any other commissioner adequately to cross-examine the officials of the Home Office or Dr. Thane, the inspector. To do this properly it is necessary that you should have studied carefully the twenty nine yearly returns that have been issued by the Home Office since the Act of 1876 was passed . . . without undue vanity I think we may claim at this office to be the sole depositories of the knowledge necessary for the elucidating the real truth."[75]

Despite his threats to ban the commission, Coleridge monitored its work, testified, and furnished it with evidence supporting the society's perspective; his letters are scattered everywhere in the commission's archives. Coleridge presented twelve charges against the Act, one of which was that the Home Office repudiated its duties under the Act and its officials being "injudicial defenders of the vivisection." Coleridge complained that the Home Secretaries' replies to the House of Commons in matters regarding the Act were "evasive and insufficient." He called for the publication of the names of those who signed vivisection certificates, as well as the name of vivisectors who violated the Act—all were missing from the annual reports. Coleridge also blamed the Home Office for failing to examine the "character of humanity" of license applicants and argued that the inspectors "displayed bias."[76]

Animal welfare and antivivisection societies submitted various kinds of responses to the commission's call, ranging from a detailed critique on the administration of the Act such as the one laid out by Coleridge, to a philosophical meditation on the legitimacy of vivisection. The RSCPA, which never opposed vivisection but insisted on pain reduction, provided its perspective that "all severely painful experiments should be carried out while the animal in question is completely under the influence of an anaesthetic, and that it should be destroyed before the effect of the anaesthetic has been removed." The word "severely" was pencil underlined in the copy kept by the Home Office. The RSCPA directed its main critique at the inspectorate and the "totally inadequate" supervision of experiments. It recommended that "all painful experiments should be prohibited except in the presence of an Inspector." Additional inspectors should therefore be appointed, "who shall not necessarily be consisted of medical men." The society also requested changes in the prosecution procedures under the Act

---

74. Secretary of the Royal Commission to Coleridge, October 31, 1906, HO 114/3.
75. Coleridge to George Wilson, November 2, 1906, HO 114/3.
76. "Final Report," 12–14.

and the abolition of the obligation to receive the Home Office approval before submitting a prosecution against a vivisector.[77]

Others submitted moral reflections. Reformer Henry Salt was among the signers of a protest letter by the Humanitarian League. Drawing from Bentham, the group compared "highly developed animals" to slaves, arguing that, "the gradual concession of a humaner treatment of animals is inseparable from a democratic society." The shared element between humans and animals was that of sentience, which science helped to uncover:

> For if there is one conceit above others that modern science has disproved, it is the old pretension that Man stands alone and distinct from the rest of organic life, and that all other beings were created for his special use; for which reason it is the more strange that physiologists should condescend to avail themselves, in the realm of ethics, of the very superstition which in the realm of science they have exposed. Having demonstrated beyond question the unity of all sentient life, they not only violate that unity by vivisection, but seek to justify themselves under the outworn pretext of a division which has no existence in fact.[78]

Home Office representatives constituted the third group of witnesses, next to antivivisectionists and scientists. The office's civil servants took a prominent part in the commission's work. They provided detailed statements about the administration of the Act and commented on the interpretation of the law. They acted as scientific experts and submitted memorandums on special issues. They also provided detailed responses to the testimony of other witnesses. The commission for example forwarded to Thane materials received from Stephen Coleridge and Lind-af-Hageby.[79]

The commission published several intermediate reports in the course of its work.[80] The final report, submitted on March 1, 1912, was composed of a brief historical background of the legislation; a summary of the Act's provisions and its implementation by the Home Office; various critiques of the Act's administration; and a response to a few alleged breaches of the Act. Additionally, the report provided an evaluation of the practice of animal experimentation, including an

---

77. RSPCA, Précis of Evidence to be given before the Royal Commission on Vivisection, November 11, 1907, HO 144/4.

78. Hebert Bell and others to Viscount Selby, Chairman of the Royal Commission on Vivisection, April 25, 1907, HO 114/4; Royal Commission on Vivisection to Henry S. Salt, May 1, 1907, HO 114/4.

79. Secretary (of the Commission) to Thane, February 15, 1911, HO 114/2; Secretary (of the Commission) to Thane, February 23, 1910, HO 114/2.

80. The *British Medical Journal* published them all.

examination of the efficacy of the practice. It also examined the question, "how far immunity from pain in experiments is or can be secured."[81]

Pain in the Second Commission

The question of pain, how to identify, measure, and control it for purposes of law—in other words how to regulate it—was a central thread in the testimonies delivered at the commission's meetings. A draft for the final report was already ready in early 1910, but its publication was postponed following a critique by CB, probably Captain Clive Bigham, Secretary to the Royal Commission, who informed chairman Ram: "On going carefully through the Report I found that there is at any rate one very important omission which I think you will agree ought to be dealt with before we send to print." CB was concerned that the report did not fulfill its mission to inquire "how far immunity from pain in experiments is or can be secured." The issue was, according to CB, "one that we cannot pass by." As he explained, "It is one of the very first things that the anti-vivisectionists will be interested in, and if we do not make a pronouncement upon it I think it will be a very serious defect in the Report." CB suggested asking commissioner John McFadyean, who had "more knowledge about the pain experienced by animals" than anyone else at the commission, to take on the task. The chapter about "immunity from pain" should deal with the topics of pithing, pain, and fish: "a quite short but explicit inquiry on these three points is all is needed."[82] He also suggested engaging with the question of allowing demonstration of experiments without anesthetics.[83]

Finally, the chapter "Pain in Experiments on Animals," examined this "most important matter" under the headings of "Anaesthetics," "Inoculation," and "Miscellaneous Questions." The latter section included "Pithing," "Experiments on Fish," and "Experiments for Demonstrations."

If the first Royal Commission approached anesthesia as a self-explanatory solution to the problem of animal pain in laboratories, the second commission was already confronted with the complexities of its use. The commission had to work through "some conflict of opinion as to the feasibility and the safety of

---

81. "Final Report," 1.
82. CB to Ram, April 9, 1910, HO 114/1.
83. CB to Ram, April 9, 1910, HO 114/1. The responsibility to write about pain immunity, pithing, and fish was later transferred to commissioner William Collins. Unsigned letter to Ram, April 12, 1910, HO 114/1.

administrating anaesthetics to animals," such as the claim, for example, that anesthesia had reduced effects on dogs. The commission examined evidence to determine "how far it is possible to subject animals used for experimental purposes to complete anaesthesia, the relative efficacy of the various agents employed for that purpose, the degree to which the disuse of anaesthetics has been sanctioned or practiced, and the extent to which the requirements of the law in regard to anaesthetics are in fact secured or may need to be amended."[84]

Inspector Thane, who witnessed ninety-seven experiments during six and a half years on duty according to his count, proclaimed he never encountered irregularities in the use of anesthetics. His testimony revealed how, by the turn of the twentieth century, anesthesia was embedded in laboratory work: "In my experience anaesthesia is practiced in an experimental laboratory as a matter of routine—not in a disparaging sense—but as an essential part of the procedure, a matter of course; in fact, anaesthesia and asepsis are carried out in the laboratory to the best of my knowledge and belief, as strictly as in the operating theatre of the hospital."[85]

Some claimed that it was not the lack of use as much as the misuse of anesthesia that was hurting animals rather than helping them, constituting a major cause of animal deaths in experiments. The surgeon Edward Laurie agreed that vivisection and anesthesia ought to be "inseparably allied" but asserted that animals suffer pain and death because of an incorrect use of anesthesia.[86] He claimed that many English physiologists erroneously perceived the greatest risk of chloroform to be heart failure, while the true risk was respiration stoppage, which led to a heart failure. In the English method, he testified, anesthesia was given through a tracheal tube. This method necessitated "a painful operation before anaesthesia is commenced" and interfered with the respiration, therefore leading to heart failure and death. The same problematic method, he pointed out, was described in *Bayliss v. Coleridge*. This phenomenon was almost unfamiliar for students of physiology in Edinburgh where a different method had prevailed. Laurie based his claim on the findings of the Hyderabad Commission on Chloroform, which, he argued, was ignored by many teachers of physiology.[87] Animal experimentation in Britain, he thus insisted, was "not done painlessly."[88]

---

84. "Final Report," 48–49.

85. G.D. Thane, Précis of Evidence, HO 114/4, 2.

86. E. Laurie, The Points I Desire to Bring Before the Royal Commission on Vivisection, HO 114/4.

87. Laurie, The Points I Desire to Bring Before the Royal Commission on Vivisection; Royal Commission on Vivisection, *Appendix to Fourth Report*, 168.

88. Laurie, The Points I Desire to Bring Before the Royal Commission on Vivisection.

Anesthesia expert Dudley Wilmot Buxton shared Laurie's doubts about the efficacy of morphine if administered alone and explained that "morphine is seldom used except with chloroform, ether or scopolamine as the anaesthetic dose acts too much upon the respiratory centre and causes death. Once injected its action cannot be abrogated. It has been largely used to lessen the quantity of chloroform employed and to lessen the muscular movements." Buxton asserted that "anaesthesia is essentially the same in man and the lower animals" and provided the commission with information about the current state of knowledge on the properties and physiological action of anesthesia. He was a past president of the Society of Anaesthesia; consulting anesthetist to the National Hospital for Paralysis and Epilepsy; and anesthetist and lecturer on anesthetic at University College Hospital. Buxton reassured the commission that pain could be scientifically tracked. By studying "the disappearance of reflexes and the ocular phenomena we are able to tell exactly how deeply a patient is under an anaesthetic. For example we can tell whether all sensation of pain is lost even though muscular movements may be present."[89]

Detecting pain was therefore a task for the professionals. Buxton was joined by "other scientific witnesses" who stressed that interpreting animal pain under anesthesia could be deceiving, and "have spoken of movements other than reflex movements, purposive movement, struggling, and vocal cries, as occurring under anaesthesia and yet not in their opinion indicative of any suffering, since in the case of man these are found on recovery to have been either unconscious or unassociated with any painful recollections." In the face of this evidence and on the background of the professionalization of anesthetics the commission dismissed doubts about the efficacy of anesthesia and concluded that with "the use of one or other or of a combination of several well-known anaesthetics complete insensibility to pain can be secured."[90]

"Cut Away the Soul of the Animal": Pithing and Decerebration

Another mechanism to turn animals insentient was the destruction of the brain, prevalent in British physiological research and teaching routine and conducted

---

89. Précis of Evidence by Dudley Wilmot Euxton, HO 114/4.
90. "Final Report," 49.

mainly on frogs and cats.[91] Common techniques included decapitation, described by inspector Russell as "the removal of the head with scissors, leaving the lower jaw, or the removal of the upper part of the head by dissection under anaesthesia"; or decerebration, the removal of the brain "by opening the skull either fully or by one or two holes and scooping out the entire brain, including the hind brain"; and pithing, which was, as one of Schafer's students explained to Russell, "the destruction of the whole brain by introducing an instrument through the occipito-atloid ligament at the back of the head."[92]

Much like with the poisonous substance curare (see chapter 1), physiologists approached the brainless body as a phenomenon to be studied, but also as a device to make the animal into a motionless, insentient, physiological structure.[93] Marshall Hall experimented with decapitating animals in the 1830s, and he was also the first to suggest that British physiologists adopt a code of practice. In chapter 1, I linked Hall's medical and normative projects. With the second Commission, the connections between the projects were no longer speculative: the question of whether a decapitated animal was insentient, or whether it was still a living being, and its proper categorization under the law were explicitly intertwined.

For many years, the operation of pithing had proceeded without the oversight of the Home Office. The procedures were conducted under a license only, and without the certificate allowing a licensee to dispense with anesthesia. Pithing basically did "not come within the operation of the Act."[94] The rationale behind the policy was that pithing transformed the animal into something the law was irrelevant to; experimenters have claimed that animals with destroyed brains should no longer be regarded as coming under the provisions of the Act, as they were "rendered permanently incapable of feeling pain."[95] Yet Inspector Russell informed the commission about a Home Office's ruling that pithing by an unlicensed person "appears to be contrary to the law," that had caused "much discussion and uncertainty among licensees and others." Physiologists asked the inspector "When is an animal legally dead in the view of the Home Office seeing that the tissues do not all die at the same time after general death; Do experiments on an isolated heart provided by an obliging butcher who killed the ox to

91. E.M. Tansey, "'The Queen Has Been Dreadfully Shocked': Aspects of Teaching Experimental Physiology Using Animals in Britain, 1876–1986," *American Journal of Physiology* 274, no. 6 (June 2, 1998): S26.

92. James Russell, "Pithing, Decerebration, Decapitation," n.d., HO 114/2.

93. Tansey, "'The Queen Has Been Dreadfully Shocked,'" S26.

94. "Final Report," 53.

95. J. McFadyean and Dr. Gaskell, "Memorandum on Pithing," draft, n.d. HO 114/1, 2.

suit the experimenter although he sold the flesh to other come under the decision." But the "chief agitation" concerned the possibility that the Home Office decision, intended for mammals, would also apply to frogs.[96]

This position resonated in Thane's *Observations*, where he contended that in the first chapter of *Shambles*, which described frogs that were pithed, had their brains destroyed, or their heads cut off, "there was no 'vivisection' at all; that is, there were no experiments on living animal calculated to cause pain." In his view, "it cannot be believed that an animal deprived of its brain can feel."[97] Concepts of life and pain often converged as those wishing to exclude pithed animals from the Act held "that the animal after being pithed was no longer 'a living animal.'" As explained by the physiologist Ernest Henry Starling, "we can cut away the higher parts of the brain, so that the animal, as an individual, no longer exists, and then we can study reflexes which, if the animal were conscious, would result from what we should call pain; we can cut away, so to speak, the soul of the animal and keep its machinery."[98]

The report's chapter on pithing was largely based on a memorandum composed by commissioners John McFadyean and Walter Holbrook Gaskell. They used human experiences to infer the lack of sensation in decapitated bodies: "There can be no doubt, from repeated observations of cases in which, as a result of accidents to human beings, the spinal cord has been severed, that sensation is entirely abolished from that portion of the body, supplied by nerves arising from the cord below the site of section." This interpretation was supported by observations on humans whose brains were destroyed by tumors, or who suffered from bleeding within their brain tissue: "From analogy from man to other vertebrates, there is, therefore, ground for the belief that in their cases also destruction of the cerebrum with the basal ganglia, and a fortiori, destruction of the whole brain would effect complete anaesthesia of the whole body."[99] Inspector Russell also submitted a memorandum, entitled "Pithing, Decerebration, Decapitation." He, too, drew from human experience in order to understand the sensation of decapitated animals:

> We only know of pain or signs of pain in other animals because we can suffer pain and give inflictions of suffering. We know that the spinal cord is not the seat of consciousness because a man with the cord interrupted by injury of disease high up feels no pain or other sensation

---

96. James Russell, "Pithing, Decerebration, Decapitation," n.d., HO 114/2.
97. Thane, *Observations*, 1.
98. "Final Report," 53, 50.
99. McFadyean and Gaskell, "Memorandum on Pithing," 1.

from anything done to the lower parts of the body . . . these experiments in man have enabled us to recognize signs of pain in the lower animals.[100]

Inspector Thane concurred with Russell's memorandum but, typically downplaying potential animal pain, added: "I cannot conceive that the head could live after the division of the upper part of the spinal cord." He explained that the shock from the operation would arrest the activity of the brain, which, deprived of arterial blood, would never be restored to activity. Furthermore, he added that "some physiologists" told him that it was possible to destroy the brain in mammals similarly to frogs, "by introduction [of] an awl through the foramen magnum of the skull," and as death was instantaneous with this method, anesthetic was unnecessary.[101] Pithing for Thane was a painless procedure to create a pain-free body.

The "painless animal" hypothesis convinced commissioners McFadyean and Gaskell, who stated in their memorandum: "After the completion of such pithing the animal in our opinion is no longer one which is subject to the provisions of the Act."[102] But this sentence was crossed out with black ink. Instead, the commission accepted the Home Office's policy and recommended that pithing would be done by licensed people, and under adequate anesthesia.[103] Other than that, the report incorporated McFadyean and Gaskell's memorandum almost word for word. It included a call for caution in interpreting the movements of decapitated animals as signs of consciousness, thus setting limits to the earlier analogy between human and animal bodies:

> In dealing with this difficult subject we are well aware of the distinction to be drawn between the consciousness of pain and the facilities whereby such consciousness may be expressed, and we realized how the inability to appreciate such distinction complicates our reasoning in such problems and warns us against the unlimited argument from analogy in drawing conclusions in regard to such questions. Thus it may be urged that even if the evidence suggests that the brainless frog displays signs of purpose and consciousness superior to those which higher vertebrate whose brain had been destroyed would exhibit, yet the susceptibility to painful influences which a fish and a frog, even with

---

100. Russell, "Pithing, Decerebration, Decapitation."
101. Thane, "Memorandum," April 23, 1911, HO 114/2.
102. McFadyean and Gaskell, "Memorandum on Pithing," 2.
103. "Final Report," 53.

the nervous system intact, can manifest appears to be of so low an order as to render it idle to apply to them such rules and precautions against possible sources of suffering as humanity prescribes in the case of the higher animals.[104]

The commission recommended that nothing other than "a complete destruction of the brain or decapitation should be accepted as equivalent to the production of complete anaesthesia," clarifying that "destruction of the brain" meant not only destruction of the cerebral hemispheres but also of the basal ganglia.[105] Pithing, despite common practice among physiologists, was relabeled by the commission as a painful procedure that had to be subjected to the supervision of the Act. At the same time the commission accepted the view that the pithed animal was equivalent to an anesthetized animal.

Pain Taxonomies

Whose pain should the Act relieve, or rather, regulate? Were all species capable of suffering pain? And within the species, were all individuals capable of experiencing pain to the same degree? Member of the Royal College of Surgeons and of the International Medical Anti-Vivisection Association, Stephen Francis Smith, arrived at the commission to protest "a widespread belief that sensibility to pain depends upon intelligence," or in other words, "that pain varies as intelligence varies." Drawing from the theory of evolution, Smith claimed that "pain has a function. The greater the sensibility to pain, the more care will the animal take to avoid danger and preserve life." Since pain was necessary for the animal to survive, its sense of pain must be highly developed.[106] Smith's view opposed the prevalent argument that tempered living conditions create delicate bodies that are prone to pain. The latter view was promoted by his fellow imperial scientists, who were committed to defining racial differences, among them, forming a racialized medical understanding of pain in which individuals in "primitive" societies were naturally tolerable to physical pain.[107]

---

104. "Final Report," 53.

105. "Final Report," 53.

106. Royal Commission on Vivisection, "Appendix to Fourth Report of the Commissioners : Minutes of Evidence, October to December, 1907" (London: printed for H.M.S.O. by Wyman & Sons, 1907), 10, 11.

107. Seth, Suman. *Difference and Disease: Medicine, Race, and the Eighteenth-Century British Empire*. Global Health Histories (Cambridge: Cambridge University Press, 2018).

Inspired by racist scientific assumptions about the intelligence of people of color in the period, Chairman Selby inquired whether the same claim was true for "different races of men," and asked if "the sensation of the negro in West Africa is more acute than that of a civilised man or as acute?" Smith replied that although the intelligence was not the same, the sense of pain was probably similar or even greater in different classes of people and animals. As chairman Selby acknowledged, Smith's evidence was a divergence from the opinion presented to the first Commission by "great many medical men."[108]

Since the Act excluded only invertebrate animals (Section 22), the Home Office decided that experiments on fish were governed by the Act.[109] In 1909, the Home Secretary asked the commission for its opinion on "experiments on fish for testing drainage affluent."[110] At that period, experiments on fish were not yet done on a large scale, and they were used mainly for water investigations. In one of the instances that the Home Office presented to the commission, the Royal Commission on Sewage Disposal asked about its use of fish in its investigations of the purity of sewage effluents into rivers: it introduced salmon fry and parr to water before and after purification treatments. The commission opined that "it well may be doubted whether such investigations were present to the minds of the authors of the Act of 1876."[111]

The question of fish pain led the commission to reconsider the species of animals protected under the law, since some witnesses called for "the omission of cold-blooded animals from the application of the Act, on the ground of their comparative insensibility to pain." This was clearly in the interest of some physiologists—the report names Schäfer, Gotch, and Horsley—who wished to remove some of the burden of applications.[112] But this path was not taken. While the commission acknowledged that "it is by no means easy to draw a line in such matters between the higher *invertebrate* and the lower *vertebrate* such as fish," it could not recommend exempting cold-blooded animals from the Act, due to "the limited knowledge which at present obtains as to the capacity of suffering in such animals." The lack of knowledge joined the need for a legal framework to direct a classification of animals: "While it would be impossible with any strict logic to define with precision the class or classes of the animal kingdom for which special legislation, in excess of the common law or of general enactments

---

108. Royal Commission on Vivisection, *Appendix to Fourth Report*, 11.
109. Royal Commission on Vivisection, *Appendix to Fourth Report*, 54.
110. W.P. Byrne to the Secretary, Royal Commission on Vivisection, June 8, 1909, HO 114/2.
111. "Final Report," 54.
112. "Final Report," 59.

against cruelty to animals, can be justified, we think there is ground for regarding with a different degree of repugnance or acceptance the employment of certain classes of animals for purposes of vivisectional experiments."[113]

The commission was "confronted with a delicate question of relative ethics" regarding the question which animals should be included in the Act:

> Here again there can be little doubt that the general moral sense of civilized mankind would be prepared to make such differentiation and would regard with quite a different degree of reprobation the like treatment for such purpose of one of the domesticated animal on the one hand with that of cold-blooded or indeed verminous or destructive animals on the other hand. The *differentia* in such case would probably be found to consist in the degree of association with or of affinity or utility to man.
>
> We feel that recognition should be accorded to the reality and worthiness of such underlying sentiment which would secure a special reservation for animals coming within the aforesaid limits. This we think that the higher apes (anthropoid) and the dog and cat present claims for special consideration.[114]

Hutton's minority report, which supplemented the report of the first Royal Commission on Vivisection, voiced the idea that the Act should treat species differently according to their intelligence and susceptibility to suffering. Hutton advocated banning the use of cats and dogs in vivisection considering the special bonds between humans and these species, as well as the increased sensibility to suffering that came along with domestication, in his terminology, "hyperaesthesia" (see chapter 2.) His opinion was considered marginal, but three decades later the second Royal Commission seriously considered "affinity or utility to man" as a criterion for classifying animals under the law, but the idea was not fully developed and was eventually withdrawn from the commission's recommendations. The commission's suggestion to provide special considerations for cats, dogs, and apes, was omitted from the report's recapitulation since it was not a "definite recommendation," and was "in effect a very slight alteration in the existing practice."[115] A previous version, which included a recommendation for the exemption of apes from experiments, was altogether withdrawn.[116]

---

113. "Final Report," 54, 63.
114. "Final Report," 57.
115. Unsigned letter to Wilson, January 30, 1912, HO 114/1.
116. Unsigned letter to Ram, May [?], 1908, HO 114/1.

The Second Royal Commission's Recommendations

Overall, the commission found that the Act fulfilled its objectives and "worked as to secure a large degree of protection to animals subject to experiment and at the same time so as not to hamper or impede research."[117] The Act established a system to which all parties have become "habituated," and which "notwithstanding its imperfections" should not "be lightly thrown away."[118] The report's recommendations focused on its administration and proposed adjustments in the implementation of the Act rather than any legislative changes. This strategy was promoted by commissioner Chalmers, a permanent undersecretary until 1908 and the Home Office representative to the commission, who commented in a memo that "if the Commission can report against legislation it is obviously desirable to do so. But a good many administrative changes which have been suggested in the course of the enquiry can be made without legislation."[119] The "Summary of Recommendations" was as follows:

1. An increase in the Inspectorate.
2. Further limitations as regards the use of curare.
3. Stricter provisions as to the definition and practice of pithing.
4. Additional restrictions regulating the painless destruction of animals which show signs of suffering after experiments.
5. A change in the method of selecting and in the constitution of the Advisory Body to the Secretary of State.
6. Special records by experimenters in certain cases.[120]

The most substantial recommended changes were related to the inspectorate and the establishment of a new advisory body. In 1910, there were 324 licensees and 48 registered places overseen by the principal inspector, 218 licensees and 43 registered places overseen by the inspector for the northern district (Scotland and six northern counties), and 23 licensees and 15 registered places in Ireland. In light of these numbers, the report recommended raising the number of inspectors to a full-time chief inspector aided by three full-time inspectors in Great Britain, in addition to one or more half-time inspectors in Ireland. It

---

117. "Final Report," 61.

118. "Final Report," 61. The report was supplemented by two reservation memorandums calling for tighter control of vivisection, one composed by Lockwood, Collins, and Wilson, and the other signed only by the latter.

119. Memo by M.D.C., January 14, 1908, HO 45/10521/138422, 1.

120. "Final Report," 64.

furthermore recommended that an inspector should be present at any experiment that involved the use of curare, in order to make sure that the animal was in a state of complete anesthesia. The physician Victor Horsley notably expressed no objection to this idea, "provided that the convenience of the investigator is safeguarded."[121]

Nevertheless, the required inspectors' qualifications remained vague. Chalmers suggested expanding the hiring criteria to include veterinary surgeons, explaining, "Certainly a trained Veterinary Surgeon as for instance a member of the Army Veterinary Service or the Indian Veterinary Service would be perfectly competent to inspect the animals in the laboratories and to direct when an animal should be killed."[122] He further commented that "although the Laboratories were numerous they are mostly collected in a few large towns and a great deal of inspection could be done by a travelling veterinary inspector."[123] The report stated that inspectors should only be qualified medical men "of such position as to secure the confidence both of their own profession and of the public."[124]

The other noteworthy recommendation was to put an end to the special arrangement that Home Secretary Harcourt had created in 1882 with the Association for the Advancement of Medicine by Research. The latter advised the Home Office regarding applications for licenses and certificates. As expected, the collaboration evoked the anger of antivivisectionists, and Coleridge characterized it as "improper private relations with a private society composed of supporters of vivisection."[125]

The Home Office under Gladstone was not pleased with Harcourt's inheritance either. Chalmers probably voiced his superior's opinion when he explained: "This is not a very satisfactory body as far as my experience goes. They merely recommend through their Secretary in all cases. Alternative body might give a much greater feeling of security to the public." The alternative body, according to Chalmers's suggestion, "might be either a small Committee appointed by the Royal Society of Medicine, or possibly by the Royal Society, or appointed by the Royal College of Physicians and Surgeons acting conjointly." He added that "it might very well be made a condition that no person holding a Licence shall be

---

121. "Final Report," 7, 61, 51.

122. Memo by M.D.C., January 14, 1908, HO 45/10521/138422, 1.

123. Memo by M.D.C. Lind-af-Hageby, who mistrusted the inspectorate, suggested allowing persons other than inspectors to be present during experiments. "Final Report," 58. Later, she supported the employment of veterinarians as inspectors. *Lind-af-Hageby v. Astor*, Third Day, 80.

124. "Final Report," 61.

125. "Final Report," 14.

a member of the Committee."[126] The report thus suggested that the home secretary would select advisers from a list of names submitted to him by the Royal Society and the Royal Colleges of Physicians and Surgeons in London. In addition, the advisers should not be holders of licenses and their names should be published.[127]

Lind-af-Hageby v. Astor (1913)

Lind-af-Hageby formed the "Miss Lind-af-Hageby's Anti-Vivisection Council," later renamed "The Animal Defence and Anti-Vivisection Society," and in June 1911 she rented a place she referred to as a 'shop' for the society on 170 Piccadilly Street, facing Bond Street.[128] The Research Defence Society, established in 1908 by Stephen Paget, employed two men to patrol outside the shop and distribute pro-vivisection leaflets and, toward the end of October 1911, it opened its own place next door.[129]

Lind-af-Hageby's shop had a large glass window exhibition, described by one of her antagonists as follows: "About 3 feet behind the said window is stretched a large canvas equal in size to the said window and completely filling up the space behind such window. Such canvas is about 6 feet high and about 8 feet long. Upon the said canvas is painted a picture which is intended to represent a vivisector preparing to operate upon a dog. At the top of the said picture are painted in large letters the words: 'Help the Helpless.'"[130]

On May 7, 1912, Caleb Williams Saleeby published an article in the *Pall Mall Gazette* condemning the Piccadilly shop. Lind-af-Hageby's first reaction was "to sit down and compose a reply," but upon rereading the article she realized that "it did not come within the category of an ordinary controversial article, and that the statement made did not come within the category of what you might call legitimate abuse," that was customary for public figures. First, she sent a letter to Saleeby demanding he publish an apology in the newspaper. Similar letters were sent to William Waldorf Astor, proprietor of the *Pall Mall Gazette*, James L.

---

126. Memo by M.D.C., 6.

127. "Final Report," 64.

128. Particulars of Justification Delivered on behalf of the Defendant Saleeby, January 10, 1913, Lind-af-Hageby Case, WA, SA/RDS/A/7, 2.

129. *Lind-af-Hageby v. Astor*, Third Day, 3. For the Research Defence Society's public action, see Boddice, *Homane Professions*.

130. Particulars of Justification Delivered on behalf of the Defendant Saleeby, January 10, 1913, Lind-af-Hageby Case, WA, SA/RDS/A/7, 3.

FIGURE 9 / Postcard of Research Defence Society shop front, with literature in the window advertising sale of Report of the Royal Commission for Vivisection, n.d. Wellcome Collection.

Garvin, the editor, and David Cameron Forrester, the printer. Yet the journal did not publish an apology. On May 10, 1912, another article appeared by Saleeby in the newspaper, along the same lines as the first. Saleeby later contended that he never received Lind-af-Hageby's first letter.[131] However, Lind-af-Hageby issued a libel lawsuit at the King's Bench Division of the High Court of Justice against him and the other three who were involved in the publication.

The court and a special jury met for fifteen successive days, excluding weekends, starting April 1, 1913. Lind-af-Hageby employed a solicitor, Shirley W. Wooler, but conducted the case at court by herself.[132] Her remarkable decision was trivialized by the *Lancet*, which when criticizing the length of the sessions, blamed it on "the consideration extended in our courts to litigants appearing in person, and such litigants not suffer if they are women."[133] Other journalists,

---

131. Particulars of Justification Delivered on behalf of the Defendant Saleeby, 42.

132. Westacott, *A Century of Vivisection and Anti-Vivisection*, 503.

133. "Lind-Af-Hageby v. Astor and Others," *Lancet* 181, no. 4678 (1913): 1176. See also Leah Leneman, "The Awakened Instinct: Vegetarianism and the Women's Suffrage Movement in Britain," *Women's History Review* 6, no. 2 (June 1, 1997): 271–87.

however, portrayed her favorably as "the new Portia."[134] The defense depicted Lind-af-Hageby as a young-though-sophisticated foreign harasser: "This lady has learned a comprehensive use of our tongue, gentlemen. She wrote that 10 years ago when she was a girl of 24 and she has stuck to ever since," contrasting her to the "practitioners of medicine and surgery of the very highest reputation" who occupied the witness box.[135] Lind-af-Hageby's opening speech lasted no less than nine hours for which Justice Bucknill, seemingly honestly, thanked her "for a very clear statement and a very nice speech."[136]

Despite some remarks to the contrary, both parties used the libel case to make grander claims about vivisection's legitimacy. The defense made explicit attempts to turn the discussion into a pro- and antivivisection debate. H. E. Duke, representing Saleeby, stated: "The issues that have been raised in this case outweigh inestimably and infinitely any of the personal questions that are involved here."[137] The defense also made numerous references to *Shambles* and called Bayliss, Waller, and Horsley—the major actors in *Bayliss v. Coleridge*—to testify. By doing so, the defense evoked the favorable decision of the *Bayliss Case*, while also linking Lind-af-Hageby's particular claims against Saleeby and the others to her broader crusade against vivisection.

For her part, Lind-af-Hageby used the trial to voice her principal charges against vivisection and to fight for the freedom of the antivivisection movement: "To me it is a very serious case: it could not be more serious. . . . I have pointed out to you that from the moral point of view, the point of view which under-lies the whole Anti-vivisection movement, the case is also a very serious one." Whatever the verdict would be, the movement will survive, because "to kill the Anti-vivisection movement would require not to prove that Anti-vivisectionists are wrong in their allegation against this scientific method or that scientific method, but to prove that the whole ethical objection is retrogressive, is something that does not go with the evolution of the world, but something that goes against the evolution of the world."[138]

Barrister Henry A. McCardie, a popular and respected lawyer, pro-vided Saleeby with a strategy: the defense should emphasize that Saleeby had made no personal attack and did not know that the shop was managed by

---

134. Cecil Cowper, ed., "Notes and News," *Academy and Literature, 1910–1914*, no. 2137 (April 19, 1913): 504.

135. *Lind-af-Hageby v. Astor*, Fifteenth Day, April 22, 1913, WA, GC/89/19, 15.

136. *Lind-af-Hageby v. Astor*, Third Day, 21.

137. *Lind-af-Hageby v. Astor*, Fifteenth Day, 2.

138. *Lind-af-Hageby v. Astor*, Fifteenth Day, 62.

Lind-af-Hageby, therefore his statements regarding the shop were not directed at her.[139] Alternatively, the defense should contend that the publication was true in substance and in fact. As Saleeby's defense planned to make use of the findings of the second Royal Commission, McCardie anticipated that Lind-af-Hageby would claim that the commission was biased. He therefore advised, "we can fairly describe Mr. George Wilson, and W. Collins and Col. Lockwood as anti-vivisectionists," and instructed the defense to further investigate Wilson's views on vivisection.[140]

In his recommended list of defense witnesses, McCardie suggested inviting no other than the Home Office inspector Thane. Next to the list the barrister added a note: "I trust that these witnesses will render every assistance. A verdict for the Plaintiff, I need scarcely point out, would be a most serious blow to the whole position of those who approve vivisection."[141] Stephen Paget thus asked Thane, in the name of the Research Defense Society, to show at the trial: "I have urgently asked to write to you. Messrs. Lewis & Lewis are very anxious to call you as a witness in the case of Lind af Hageby v. the Pall Mall Gazette. I told them that I have no 'influence' with you. But, of course, I do very earnestly hope that you will not refuse to give evidence. For the case is one of great importance to the public."[142]

Thane had to decline the request. He was bound by a memorandum prepared by Digby in November 1896, instructing inspectors not to appear in civil suits as witnesses unless subpoenaed, and commanding that inspectors "should not communicate with or supply proofs to either of the parties. In exceptional cases, where it may be necessary to give such information, it should be given to *both* parties."[143] In light of these limitations, Paget advised Saleeby's lawyers to withdraw the idea of summoning Thane to the witness box.[144]

The attempt to call an inspector to testify in a civil lawsuit revealed how central the Act's administration was not only to setting the norms for physiological research but also as a source of information about scientific practice. Although it was a tort case and not a criminal prosecution under the Vivisection Act, the parties, and in particular the defense, used the Act repeatedly in their arguments.

---

139. Henry A. McCardie, "Advice on Evidence to the Defendant Saleeby, Lind-af-Hageby Case," WA, SA/RDS/A/7.

140. McCardie, "Advice on Evidence," 6.

141. McCardie, "Advice on Evidence," 13.

142. Paget to Thane, January 25, 1913, HO 144/1250/233716.

143. Kenelm Digby, "Instruction as to the Attendance of Home Office Inspectors as Witnesses in Civil Cases," November 10, 1896, HO 144/1250/233716.

144. Paget to Thane, January 30, 1913, HO 144/1250/233716.

In a document prepared by Saleeby's defense team, the shop was described as misguiding the public because it failed to acknowledge the Act and its restrictions over vivisection: "None of the exhibits in the said window suggest and in particular the said card describing the wooden operating board carefully omits to indicate that all experiments upon animals in this country are restricted by the safeguards and precautions" imposed by the Act.[145]

An attack on vivisection was, therefore, an attack on English law: "This window seems to me to have this underlying vice of false statement, that it presents to the English public in this agitation addressed against English law and English practice, representations of fact which could not lawfully come into existence in this country but which are facts which could only lawfully come into existence in France or Germany or some country where there are not the restrictions there are here." Saleeby's solicitors further contended that none of the exhibits in the shop's window "in any way indicate that by reason of the said Act of 1876 and the practice of the Home Office thereunder, no such operations may be performed in this country unless the animal is completely anaesthetized, that is, wholly insensible to pain, during the whole of such operation." The shop's exhibition thus made the deceptive suggestion that "vivisectors in this country, that is, students of physiology and scientists engaged in physiological research, habitually infringe the provisions of the said Act."[146] In the hands of the defense, the Act became descriptive rather than prescriptive.

In addition to misrepresenting British law, the defense argued that Lind-af-Hageby was carelessly harming the Home Office's reputation: "The Act had been administrated by the Home Office by ten successive Secretaries of State everyone of whom the Plaintiff has thought fit to charge with indifference to his duty in the protection of the lower creation under the Act." The defense emphasized that in thirty years of the Act, "the solitary incident which has been raised was the scandal raised at the instance of the Plaintiff herself in 1903," referring, again, to the *Bayliss Case*.[147]

The Home Office's bureaucracy provided plentiful evidence. Barrister McCardie, who prepared the defense strategy, instructed the litigators to bring to the court "Annual Reports of the Home Office inspector for a number of years. Also have in court examples of all the licenses and certificates with the Home Office conditions thereon which are issued; they should show that restrictions

---

145. Particulars of Justification, 5.
146. *Lind-af-Hageby v. Astor*, Fifteenth Day, 59.
147. *Lind-af-Hageby v. Astor*, Fifteenth Day, 11.

are places upon 'vivisection.'"[148] Based on the published Home Office reports, the defense emphasized that "about 95 per cent of all experiments under the Act are either inoculation experiments the results of which are in the vast majority of cases either negative or painless, or are experiments which do not involve inoculation but are not calculated to give any serious pain, and therefore about 95 per cent of all experiments under the Act are performed under Certificate 'A.'"[149] To recap chapter 3, the pain involved in inoculation experiments was debated at the Home Office, and the subsequent question of the legal classification of disease research was handed to the Law Office several times. It was decided that experimenters should hold Certificate A, which was mandatory for painful experiments without anesthetics, for inoculation experiments. The defense attempts to revoke the old doubts about the pain involved in inoculation.

A Glass Eye's Agony

Three depictions of dogs at the antivivisection shop's window became central to the dispute. The first was a painting of a dog stretched out upon an operating board, "its legs are tightly trenched out and securely tied to each of the four corners of such board. Its head is shewn as rigidly fixed by a metal clamp secured tightly down upon the top of its muzzle." On a shelf nearby, "a large and coarse-toothed saw which it is intended to suggest is about to be used upon the said dog for the purpose of vivisecting it." But most important, "the said dog is painted with its eyes wide open, and is so painted as to indicate clearly that it is either in a severe agony or is in a state of painful terror." A second exhibit at the window was a wooden board, upon which laid "a brown dog of the terrier type the body of which has been stuffed and set up in a life-like manner." The dog was tied down, and his glass eyes were wide open. A third exhibit at the window was a poster, titled "The Effect of Publicity on the Vivisector," depicting a vivisector operating on a small dog, "shown as being disturbed by the sudden turning of a search-light upon him."[150]

Now, not only living bodies were scrutinized for signs of pain, but also animal images. The defense used the display depictions of opened-eyed dogs to claim that the antivivisectionists distorted and misrepresented reality: How could the eyes be wide open, if the Act required vivisectors to put animals under

---

148. McCardie, "Advice on Evidence," 5.
149. Particulars of Justification, 7, note G(1).
150. Particulars of Justification, 4.

anesthesia? Saleeby's lawyers argued that none of the exhibits in the window "in any way indicate that by reason of the said Act of 1876 and the practice of the Home Office thereunder, no such operations may be performed in this country unless the animal is completely anaesthetised, that is, wholly insensible to pain, during the whole of such operation."[151] Along the same lines, Montague Shearman, who represented defenders Astor, Forrester, and Garvin, contended: "It is demoralising exhibition because it is unscrupulously says that it is accusing the system of cutting up a dog which is obviously alive and conscious, when, as a matter of fact, it is untruthful representation" since "the English system is these days is never to cut up a conscious animal at all."[152]

This argument should not have surprised Lind-af-Hageby. Two years earlier, Stephen Paget, honorary secretary for the Research Defence Society, criticized a photo shown in Lind-af-Hageby's *Antivivisection Review*, in a letter in the *Saturday Review*. The photo showed a vivisected dog, "But, in the picture, there is no one word about anaesthetics." Paget encountered the same in her shop window, where there was "a stuffed animal tied on a board; but there was not one word to say, or even to suggest, that no operation on any animal, more than the lancing of a superficial vein, is allowed in this country, unless the animal is under an anesthetic throughout the whole of the operation." The antivivisection shop, therefore, "did not tell, but hid the truth about experiments on animals in this country," because it failed to reflect the legal requirements.[153]

Saleeby's lawyer asked one of the illustrations' painters, William Luker, about the opened-eye dog: "Did you intend to convey to the public who saw your picture that although the law required an anaesthetic the operator might dispense with it?" The painter replied he was influenced by prints he had seen, and did not think of "representing any particular thing" but also commented that he heard it was "quite possible" that some experiments were done without the animal being anesthetized.[154]

Lind-af-Hageby maintained that the posture of the stuffed dog in the display drew from an image in a scientific instruments catalog. She explained that the stuffed dog was "simply there to show how animals are strapped and gagged. They say the eyes are open. Yes, the eyes are open, but the eyes are open not because I went and said 'Make the eyes open.' Gentlemen, I never thought of

---

151. Particulars of Justification, 5.

152. *Lind-af-Hageby v. Astor*, April 21, 1913, WA, GC/89/18, 4, 6.

153. Stephen Paget, "Anti-Vivisection Shops," *Saturday Review of Politics, Literature, Science and Art* 112, no. 2907 (Jul 15, 1911): 82, https://search.proquest.com/historical-periodicals/anti-vivisec tion-shops/docview/876905147/se-2?accountid=14765.

154. *Lind-af-Hageby v. Astor*, Tenth Day, April 14, 1913, WA, GC/ 89/12, 99–100.

that point. I took the trade catalogue, that page which you saw, to Rowland Ward, and I said 'Do me this dog exactly like the dog in this catalogue.' That is why the eyes are open." Furthermore, Lind-af-Hageby used the catalog image to make a countering point—that the need for strapping testified that vivisectors were not using deep anesthesia, adding that "many experiments could not be successful if the animal were in a state of absolute anaesthesia."[155]

A similar discussion about pain hermeneutics in antivivisection images evolved at the court around a cartoon in a leaflet that Lind-af-Hageby had published in 1909, described by the Justice as the following: "The dog is looking right round; it looks as if it is not under an anaesthetic at all. This is a dog which the man has picked out and put upon a table, and just as he is about to operate with a knife light is thrown upon it, he drops the knife, and assumes the expression which is in the picture; the dog is looking round, and a dog below is looking up."[156] Justice Bucknill asked Lind-af-Hageby whether the image was "supposed to represent an operation where the man is going to out the dog whilst the dog is wide awake?" But she rejected the attempt to make her representations into a concrete event with relevance to the Act: "It may or may not be. That cartoon does not refer to any particular class of experiment or to the Act . . . It never occurred to me at the time that anyone could put this literal interpretation upon it which is put to me today. The face of the man is enormous, the lantern shows it is not a real picture of a hand. I never thought of it from the point of view of the law."[157]

Lind-af-Hageby had no belief in the Act. She cultivated some interest in the inner physiological discourse when she was a young medical student who wished to argue with the profession from within. The *Shambles*, she admitted in the court, did not contain any reference to the Act as she had no interest in the legal context of physiological research when she composed it. In her words, "at the time when I wrote that book I knew very little indeed about the Act of 1876 and indeed my whole interest and my whole attention were concentrated upon vivisection, physiology and theories on physiology rather than upon the legal aspect of the question."[158] However, during the decade between the *Bayliss Case* and her lawsuit against the *Pall Mall Gazette*, "I have concentrated deliberately on studying the Act of 1876 and have gone fully into it."[159]

---

155. *Lind-af-Hageby v. Astor*, Third Day, 15.
156. *Lind-af-Hageby v. Astor*, Fifth Day 2nd Part, April 7, 1913, WA, GC/89/4, 60–61.
157. *Lind-af-Hageby v. Astor*, Third Day, 62.
158. *Lind-af-Hageby v. Astor*, Third Day, 73.
159. *Lind-af-Hageby v. Astor*, Third Day, 73.

But knowing the Act and its implementation policy at the Home Office did not mean softening Lind-af-Hageby's critique of vivisection. For example, she contended that certificate B, which allowed keeping an animal alive after experiments under the condition that it would be killed if it experienced severe suffering, was "not in fact acted upon. I say that as a fact animals are kept alive in a state of severe and terrible suffering and I say there is nothing in the law at present to prevent a vivisector keeping animals alive for weeks and months if in his view he has not attained his object." The Home Office condition was "practically a dead letter, a pious opinion and nothing more."[160]

Lind-af-Hageby had lost in the trial, but, as in the *Bayliss Case* with which she was indirectly involved, she obtained valuable publicity for the antivivisection cause.[161] In two months, her supporters raised a sum of £6,000 to cover her trial costs, contributions ranging from one penny to £1,000.[162] The shop remained open until May 1913, attracting hundreds of thousands of visitors.[163]

Coming to terms with the Act in its early days, scientists explained that good science was also moral science. A pain-free animal would be morally justifiable and produce reliable scientific facts. Lind-af-Hageby also believed in the inevitable overlap of good and right, but the failures of the law had driven her to the conclusion that vivisection should be terminated: "For right at the bottom of my heart, and the bottom of my soul, there is a profound conviction that that which is morally wrong, spiritually retrogressive, cannot in the long run be scientifically right, and I believe that if we wait ten years, twenty years, thirty years, hundreds of years, it will be found that that which is spiritually right, spiritually beautiful, spiritually good, will also in the long run be found to have been physically useful."[164]

The Vivisection Act, however failed in Lind-af-Hageby's eyes, was an embodiment of the vision of the convergence of ethical conduct and sound science, in other words, the merger of the morally and scientifically right. The same logic was presented in an appeal discussed by King's Bench Division in 1914. The defendant, the parasitologist Warrington Yorke, and the director of a research laboratory in connection to the Liverpool School of Tropical Medicine, strikingly

---

160. *Lind-af-Hageby v. Astor*, Third Day, 75.

161. "Medico-Legal," *British Medical Journal* 1, no. 2730 (1913): 917–18.

162. Westacott, *A Century of Vivisection and Anti-Vivisection*, 505.

163. It is unclear whether Lind-af-Hageby closed the shop following the court decision. Westacott, *A Century of Vivisection and Anti-Vivisection*, 502.

164. *Lind-af-Hageby v. Astor*, Third Day, 21.

commented that "the value of the animal from a standpoint of scientific research was a guarantee that it was treated with every consideration."[165]

The Act's presence in private actions concerning the reputations of scientists and antivivisectionists shows how central it was in framing the ethical issues raised by subjecting animals to experimentation. By the turn of the twentieth century, the Home Office's civil servants were the most authoritative resource for knowledge about British biomedical research in general and animal pain in particular. The Act itself became evidence for the research routine in physiological laboratories. Above all, scientists used the Act as an indication for the use of anesthetics. Some interpretations of suffering were dismissed as misjudgments, as according to Thane, "The look of utmost agony in the eye is the imagination of the writer." At the same time, the presentation of the painful vivisected animal conflicted with the law, even when the said animal was a stuffed dog in a window display.

---

165. Dee v. Yorke (1914), 12 LGR 1314, 78 JP 359, 30 TLR 552, 553. RSPCA inspector Harold Leach Dee filed a summons against Yorke, contending he had violated section 1 in the Protection of Animals Act (1911) when he unlawfully and cruelly caused unnecessary suffering to an ass by failing to give it the proper care and attention while it was in a suffering state. The case was dismissed.

CONCLUSION / The Act in the Twentieth Century

The history of vivisection regulation shows the crucial role law played in producing physiological knowledge, starting in the late nineteenth century. The Vivisection Act (1876) shaped experimental settings and routines: it influenced the design of space, the planning of experiments, and the keeping of records. By 2001, a license could reach as many as three hundred pages.[1] The number of inspectors in the UK culminated at twenty-five, making 2,100 site visits each year.[2] At the same time, scientific concepts and methods influenced the legal framing of ethical conduct in laboratories. Physiologists were central to providing policymakers with the latest knowledge about the body's sentient capacities, anesthesia technologies, and other ways to manipulate animal pain.

This dynamic, which structured the first three decades of the Act's implementation, continued to characterize British animal experimentation laws throughout the twentieth century. The legal regime established by the Act obligated state administrators and scientists alike to empathize with animals in the sense of understanding nonhuman pain in order to comply with legal demands. Enforcing the law was therefore entangled with an experience of affect, and recognizing animal pain was part of the bureaucratic procedure. New formulates for decision-making about the condition of animals in research were designed to refine the bureaucracy of empathy and provide both civil servants and scientists with elaborated models to interpret and classify animal pain. The attempts to set rules and guidelines for detection and evaluating animal pain grew sophisticated and, later in the century, transnational. The scope of relevant expertise widened to include not only physiology but also animal behavior and psychology. During the twentieth century, identifying and quantifying pain moved from being an

---

1. Report of the Select Committee on Animals in Scientific Procedures, HL Paper 150, Session 2001-02, paragraphs 5.39–5.40.
2. Report of the Select Committee on Animals in Scientific Procedures, paragraphs 5.5 and 5.8. The Select Committee recommended the reduction of a typical license length to ten pages.

implicit task in implementing the law to a central component of vivisection legislation.

The Littlewood Report (1965)

The Vivisection Act remained substantially unaltered for more than a century, while Britain's number of experimental animals surged to millions.[3] In the two decades following the second Royal Commission, antivivisection societies promoted various unsuccessful bills to replace the Act. The prohibition on police officers giving or selling seized stray dogs for vivisection under the Dogs Act 1906 was a single achievement.[4] Notwithstanding this stagnation, other aspects of animal welfare legislation evolved. For example, the British Parliament consolidated all previous cruelty to animals laws under the Protection of Animals Act (1911), explicitly excluding operations lawfully done under the Vivisection Act.[5]

The attempts to amend or replace the Act stopped in the thirties, and public concerns with animals in laboratories continued in a lower key.[6] At the same time, standards for the treatment of these animals developed within the scientific community. In 1947, the Medical Research Council established the Laboratory Animals Bureau (later to be called the Laboratory Animals Centre) to examine the procurement and husbandry of laboratory animals.[7] In 1959, the zoologist and psychologist W. M. S. Russell and microbiologist R. L. Burch spelled out the three R's principle, which became the leading imperative in vivisection law: replacement of "conscious living vertebrate" with "non-sentient material" whenever possible; reduction in the number of animals used in research; and refinement of the experimental procedures.[8]

---

3. By 1948, over 1.5 million animals were used in experiments, and it increased constantly. Sydney Littlewood et al., "Report of the Departmental Committee on Experiments on Animals" (London: Her Majesty's Stationery Office, 1965), 13, 55.

4. Dogs Act 1906, c.32, Subsection 3(5).

5. Subsection 1(3).

6. Littlewood Report, 1, 73–76.

7. Littlewood Report, 19. For more on the Laboratory Animals Bureau and about the development of a new discipline of "laboratory animal science," see Robert G. W. Kirk, "A Brave New Animal for a Brave New World: The British Laboratory Animals Bureau and the Constitution of International Standards of Laboratory Animal Production and Use, circa 1947–1968," *Isis* 101, no. 1 (March 2010): 62–94.

8. W. M. S. Russell and R. L. Burch, *The Principles of Humane Experimental Technique*, special ed. (South Mimms, Potters Bar: Universities Federation for Animal Welfare, 1992), http://altweb.jhsph. edu/pubs/books/humane_exp/het-toc.

In the early sixties, the Royal Society for the Prevention of Cruelty to Animals (RSPCA) revived the critique of the administration of the Act.[9] On May 23, 1963, Home Secretary Henry Brooke initiated a reevaluation of the Act by the Departmental Committee on Experiments on Animals, chaired by Sir Sidney Littlewood.[10] Unlike the first and the second royal commissions on vivisection, the Littlewood Committee visited experimental facilities and observed animals before, during, and after experimentation. The home secretary presented the Littlewood Report to the Parliament two years after its completion. The report addressed the rising number of experiments, explaining it by the emergence of new branches in biological science, its growing complexities, and moreover, "the need for quantitative assessments based on animal tests." Safety and efficacy tests were mandated by law, and statistical analysis demanded many animals.[11]

In the decades since the ·Act's legislation, surgical experiments lost their prominence in British science. Vivisection "was no longer an accurate description of animal experiments as a whole." The committee favored using the word 'experiment' to mean "an experimental procedure carried out on a living animal and 'acute experiment' to mean a surgical procedure upon an animal made unconscious and not allowed to recover."[12] The report expected a future increase in the use of animals in fields such as immunology, endocrinology, genetics, and nutrition, where anesthesia was not required.[13]

The changes in research agendas, the Littlewood Report advised, should be addressed with updated legislation: "The growth of experiments in which the animal spends most of its life in an animal house with intermittent visits to the laboratory will make such matters as housing, feeding and general care of animals, of growing humane, scientific and economic importance."[14] Additionally, the report called on lawmakers to address neglected issues such as animal supply, killing methods, and the pain of genetically modified animals.

The Littlewood Report was an important milestone in the creation of the "laboratory animal." By inquiring into the supply of animals for experiments, the committee responded to public dread of pet stealing and illicit trade.[15] This

---

9. Littlewood Report, 13–14; 659 Parl. Deb., H.C. (5th ser.) (1962) 689.

10. Littlewood Report, 1. See also Statement by the Home Secretary on November 30, 1962, Littlewood Report, Appendix III, 211. Littlewood Report, 1. About the public pressure preceding the appointment of the Littlewood Committee, see Robert Garner, *Animals, Politics and Morality* (Manchester: Manchester University Press, 2004), 13, 181.

11. Littlewood Report, 1, 24.

12. Littlewood Report, 2.

13. Littlewood Report, 26.

14. Littlewood Report, 26.

15. Littlewood Report, 173, 178–79; 724 Parl. Deb., H.C. (5th ser.) (1966) 1796.

century-old anxiety was not ungrounded, as many researchers confirmed their dependency on animal dealers to obtain cats and dogs for research. From their perspective, the use of stray dogs and cats was not ideal either, as the supply was unreliable, and the animals were susceptible to infection.[16] The concept of special breeding units was accepted, even welcomed, by both experimental bodies and animal welfare organizations showing at the committee.[17] At that time, the commercial breeding of smaller mammals for science was already underway. In 1950, the Laboratory Animals Centre published a voluntary accreditation scheme for breeders of mice, guinea pigs, and rabbits. Both the economic aspects and emotional concerns led the committee to the uncontroversial conclusion that rather than using stray cats and dogs, "the aim should be to breed dogs and cats as well as the conventional laboratory animals specially for experimental use."[18]

Littlewood's Unfolding of Discomfort, Stress, and Pain

The Littlewood's commission "quickly became aware that the concept of pain is complicated and difficult to define. Most people are aware from personal experience that there is no standard terminology for describing sensations or degrees of suffering; and that it is as difficult to communicate their own feeling as it is to judge those of others." The commission acknowledged that some human pain might be measurable by physical indices, "the technique, however, is costly and elaborate and in the foreseeable future it seems that the assessment of pain must remain largely subjective."[19]

The uncertainty in defining pain ("does it or does it not include malaise, discomfort, frustration and fear?") was "a cause of anxiety to many licensees."[20] Many scientists testifying in the committee suggested replacing the term "pain" with "any interference with or departure from the animal's normal state of health or well-being," a state that can also be described as "discomfort" or "distress."[21]

---

16. Littlewood Report, 175–76. Tansey discusses practices and problems of animal supply in her study on the Department of Physiology of the University of the Bristol laboratory. E. M. Tansey, "'The Queen Has Been Dreadfully Shocked': Aspects of Teaching Experimental Physiology Using Animals in Britain, 1876–1986," *American Journal of Physiology* 274, no. 6 (June 2, 1998): S27. In the US context, see Adele E. Clarke, "Research Materials and Reproductive Science in the United States, 1910–1940," in *Physiology in the American Context, 1850–1940*, ed. Gerald L. Geison (New York: Springer, 1987), 323–50.

17. Littlewood Report, 181–84.

18. Littlewood Report, 174, 183.

19. Littlewood Report, 54.

20. Littlewood Report, 55.

21. Littlewood Report, 56.

Admitting that it was "not as a rule possible to assess degrees of real pain in animals," nearly all witnesses agreed that there was "no scientific evidence that any single species was more sensitive to pain than another."[22] Half a century earlier, the common view was that species experienced pain differently, and that difference also existed between wild and domesticated animals.

Despite pain being a subjective symptom, and its detection "in the inarticulate animal" presented "formidable problems," the committee stressed that legislators were responsible for guidance regarding the proper approach to the problem of pain. The report advocated a refined definition of pain and proposed to recognize three states of suffering: discomfort, "such as may be characterized by such negative signs as poor condition, torpor, diminished appetite"; stress, "i.e. a condition of tension or anxiety predictable or readily explicable from environmental causes whether distinct from or including physical causes"; and pain, "recognizable by more positive signs such as struggling, screaming or squealing, convulsions, severe palpitation."[23]

The committee concluded that the Act should be profoundly reformed. It provided eighty-three recommendations for administrative action and the introduction of new legislation, among which was a call to clarify and broaden the concept of pain: "The Act should be amended so as to clearly apply to any experimental procedure liable to cause pain, stress, interference with or departure from an animal's normal condition of well-being."[24] It also recommended imposing the law not only on the experimental procedure but also over the whole life of animals held in laboratories as well as to "animals born of an animal under experiment."[25] This administrative oversight should continue to the animal's death—the committee recommended that "guidance should be given to licensees on methods of painless killing." Future legislation should apply to "animals used for production of biological products." Breeding of laboratory animals with susceptibility to spontaneous disease should be considered an experimental procedure, and breeding of animals for research should be brought under statutory control.[26] Other topics included revising the license system, control over premises and husbandry, and the inspectorate.[27]

---

22. Littlewood Report, 59, 60. This was embraced as one of the committee's findings. Littlewood Report, 190.

23. Littlewood Report, 56, 57.

24. Littlewood Report, 190.

25. Littlewood Report, 191, 192.

26. Littlewood Report, 194, 192.

27. Among others, the committee recommended that candidates for the inspectorate should be between thirty-five to forty-five years old, and that inspectors' salaries should be raised. Littlewood Report, 193, 195, 196.

The Littlewood Report was not debated in the House until six years after its publication.[28] It took another two years before the next debate on the report took place due to parallel legislative attempts at the European Union. Criticizing the delay, MP Douglas Houghton contended that "the laboratory doors are closed by the 1876 Act. The Littlewood Report recommended that those doors should be opened."[29] In the following years, the Home Office implemented several of the report's recommendations that did not require parliamentary approval. For example, it appointed lay members to the Home Office Advisory Committee (which had been established in 1913 following the recommendations of the second Royal Commission) and reorganized the inspectorate with the appointment of supervising inspectors and an increase in the number of inspectors with veterinary qualifications.[30]

The House of Lords Select Committee (1980)

In 1979, two private members' bills to replace the 1876 Act were introduced, incorporating the Littlewood Report's recommendations.[31] MP Peter Fry introduced the Protection of Animals (Scientific Purposes) Bill in the House of Commons on June 27, 1979. Fry prepared his bill with the assistance of RSPCA officials, but facing strong opposition from groups associated with the medical lobby; he withdrew it on June 4, 1980 at the Standing Committee stage. Earl of Halsbury introduced the Laboratory Animals Protection Bill, composed with the assistance of the Research Defense Society, to the House of Lords on July 16, 1979.[32] On October 25, 1979, the parliament gave a second reading to the Laboratory Animals Protection Bill. It was then referred to a Select Committee of the House of Lords.[33]

---

28. 818 Parl. Deb., H.C. (5th ser.) (1971) 1395.

29. 856 Parl. Deb., H.C. (5th ser.) (1973) 887.

30. Memorandum Submitted by the Home Office, December 13, 1979, in House of Lords, *Report of the Select Committee on the Laboratory Animals Protection Bill*, vol. 2 (London: H.M.S.O, 1980), 12. In 1979, two private members' bills to replace the 1876 Act were introduced and both incorporated the recommendations of the Littlewood Report. House of Lords, *Report of the Select Committee on the Laboratory Animals Protection Bill*, vol. 1 (London: H.M.S.O, 1980), 246-I, 27.

31. House of Lords, *Report of the Select Committee on the Laboratory Animals Protection Bill*, vol. 1 (London: H.M.S.O, 1980), 246-I, 27.

32. Advisory Committee on Animal Experiments, Report to the Secretary of State on the Framework of Legislation to Replace the British Cruelty to Animals Act 1876 (London, 1981), 7.

33. *Report of the Select Committee on the Laboratory Animals Protection Bill*, 16.

The Select Committee on the Laboratory Animals Protection Bill published its report and an amended bill in April 1980. Among the many organizations to submit evidence to the committee was the Association for the Study of Animal Behaviour. The association's president, Patrick Bateson, opined that the list of protected species should be extended to include "all those that current neurological and behavioural knowledge suggest have complex nervous systems and are sentient. This applies particularly to long-lived animals such as the *Octopus*. Therefore, the list should at this stage include cephalopods and the late embryos of birds and reptiles and not exclude the larval form of amphibian." Bateson also claimed it was essential to add psychologists to the composition of the Advisory Committee "since the assessment of pain, sensation, stress, and so forth relies so much on behavioural evidence, and since psychologists use so many animals."[34]

The British Psychological Society joined Bateson's call to expand the Act to oversee also psychological research. The society emphasized that psychologists were "uniquely well placed to make judgments concerning stress to which animals may be exposed, since the concept of stress is at least in part psychological" and called for the inclusion of a psychologist in the Advisory Committee. Furthermore, it required that the amended law "include psychological procedures which may involve stress to animals clearly and unambiguously" within its terms.[35]

There was an urgent need for new legislation. The Vivisection Act was outmoded "for modern biology has overtaken the terms of the Act as dramatically as the motor car has outdated the road traffic laws of the 1870s." The committee drafted a modified version of the Laboratory Animals Protection Bill, offering "a more rational licensing procedure and policy."[36]

However, the legislation was delayed since, at the same time, the European Council was working on a convention for the protection of vertebrate animals used for experimental and other scientific purposes.[37] On June 20, 1980, Earl of Halsbury, who drafted the Laboratory Animal Protection Bill, recommitted the amended bill and attempted to convince the House of Lords not to wait for the European Convention before replacing the existing Act. He said he had sent his bill to the convention secretariat in Strasbourg and emphasized "the

---

34. Letter from the President of the Association for the Study of Animal Behaviour, February 21, 1980, *Report of the Select Committee*, 2:242.

35. Letter from the British Psychological Society, December 4, 1979, *Select Committee on the Laboratory Animals Protection Bill*, 2:250.

36. *Report of the Select Committee on the Laboratory Animals Protection Bill*, 25, 3–15, 35.

37. Advisory Committee on Animal Experiments, Report to the Secretary of State, 9.

importance and the need for Britain to show leadership in this field."[38] In a debate a few months later, Earl of Halsbury criticized the existing Act for being loosely defined, thus providing the Home Office with "a monopoly of authority" of which "the public knows very little in the matter of detail."[39] Again, he stressed that waiting for Europe was undermining Britain's pioneering position on animal welfare issues:

> I wanted us to give Strasbourg a lead by introducing legislation they could copy into their convention based on our 100 years of experience in this field. I do not believe they would have been the least bit adverse to our doing so, for I have visited Strasbourg and talked to the officials engaged on the task and they are well aware that lawyers as such are quite out of their depth in this sphere and that what is needful is experience.[40]

The Advisory Committee's Report (1981): Redefining Protected Species

In May 1980, against the background of a growing agitation and with bills to replace the Vivisection Act piling up, Home Secretary William Whitelaw instructed the newly composed Advisory Committee on Animal Experiments to study the proposals before the Houses of Parliament and the Council of Europe. The European document was yet to be published. Still, the commissioners were well aware of its prospective principles since the UK delegation, the head of which was the chairman of the ad hoc committee drafting the convention, kept them informed.[41]

The committee met ten times and set up several subcommittees. It did not take evidence from organizations or individuals, a decision explained by time constraints and the fact that there was already "extensive literature on the subject." Yet, it obtained additional information on specific matters. The Advisory Committee was satisfied that it knew "the views of most of those involved in, and concerned about, the operation of the 1876 Act." Clarifying that its function was not "to arbitrate between the Bills," the committee had attempted to "formulate general principles upon which new legislation should be based, taking into

---

38. 410 Parl. Deb., H.L (1980) (5th ser.) 1323.
39. 415 Parl. Deb., H.L (1980) (5th ser.) 1250.
40. 415 Parl. Deb., H.L (1980) (5th ser.) 1252.
41. Advisory Committee on Animal Experiments, Report to the Secretary of State, 1, 10.

account the deficiencies of the 1876 Act and the different approaches adopted by those who drafted the Bills."[42]

The report, published in May 1981, concluded with twenty-five recommendations. One of the most substantial of which was the redefinition of protected species, prioritizing the species' capacity to feel pain. The committee inquired into the possible inclusion or exclusion of fetal, embryonic, or nonreproducing larval forms. In particular, it studied "whether or not there is evidence that embryonic forms of fish are capable of experiencing pain." The views were "unanimous that there is no scientific evidence that fish embryos experience pain as it is generally understood; but they were equally unanimous in their view that fish embryos do react to noxious stimuli (though this need not involve brain centers and might involve only spinal reflexes)." Most of those who consulted the committee on the topic "preferred to err on the side of caution, to assume that pain could be experienced." Additionally, "because of the unequal rates of maturation in the various species and the difficulties in deciding exactly when the larval form develops its nervous system, it was suggested to us that protection should be given from the time of the hatching of the egg."[43]

The report concluded that deciding upon which life stages should be included under the Act was "a particularly difficult area, in which the evidence is not at all clear-cut." The new legislation "should extend to all animals of the species of the Sub-phylum Vertebrata of the Phylum Chordata including the foetuses of mammals, and the embryonic or larval young of members of other Classes (whether oviparous or viviparous), that have attained such a stage of development that they are capable of a discrete existence outside of the egg or maternal tract." The Advisory Committee expressed its will "to see flexibility in legislation to enable protection to be extended in the future to an even wider range of sensate creatures." In the opinion of some commissioners, "it may even now be appropriate to protect certain non-vertebrate forms, for example octopuses and crustacea."[44]

The Advisory Committee challenged the special status under the existing law of cats, dogs, and Equidae, claiming that "there is no reason to suppose that these animals are more sensitive to suffering than others." The committee acknowledged public feelings and recognized "that these are companion animals who stand in a special relationship of trust with humans." However, it aimed to "raise the level of protection to all species" and argued that this would not be achieved "by legislative provision for particular species to be treated differently

42. Advisory Committee on Animal Experiments, Report to the Secretary of State, 2, 9.
43. Advisory Committee on Animal Experiments, Report to the Secretary of State, 11.
44. Advisory Committee on Animal Experiments, Report to the Secretary of State, 11, 12.

from others." Nevertheless, the committee also recommended affording "special protection" to primates not by law but "by administrative means."[45]

The Animals (Scientific Procedures) Act 1986

The Vivisection Act was finally repealed and replaced on May 20, 1986 by the Animals (Scientific Procedures) Act 1986.[46] The changes in the law reflected the experience of the Home Office while implementing the old Act. While the 1876 Act regulated "any experiment calculated to give pain," the 1986 Act referred to "any experiment or scientific procedure applied to a protected animal which may have the effect of causing that animal pain, suffering, distress or lasting harm."[47] The addition of the term "scientific procedure" aimed probably to minimize the complexities in defining an experiment, as demonstrated in the case of inoculation operations (see chapter 3). The "lasting harm" addressed the ambiguities in situations such as those in Ferrier's brain experimentation, where the pain was not clearly manifested (see chapter 2). Additionally, section 14(1) of the new act banned the reuse of protected animals in research, a lesson probably taken from the *Bayliss Case* (see chapter 5).

The new act incorporated many of the policies developed by the Home Office. The appointment of inspectors with medical training became explicit with the instruction to appoint persons "having such medical or veterinary qualifications" as the home secretary thought requisite. While formerly the inspectors were formally only responsible for visiting registered places, Section 18 of the new Act spelled out, in addition to visiting, the duty to advise the Home Secretary on applications for licenses and certificates, to review the applications, to report any deviations, and to advise the home secretary on proper responses.

Furthermore, the 1986 Act allowed the inspector, if he "considers that a protected animal is undergoing excessive suffering," to require it to be immediately killed.[48] The latter incorporated a license condition (the "pain condition") initiated by inspector John Eric Erichsen in 1887 and refined by the second Royal Commission.[49] In addition, the 1986 Act required the establishment of

---

45. Advisory Committee on Animal Experiments, Report to the Secretary of State, 30, 23, 35.
46. Animals (Scientific Procedures) Act, 1986, c. 14.
47. Animals (Scientific Procedures) Act, Subsection 2(1).
48. Animals (Scientific Procedures) Act, Section 18.
49. See chapter 3.

a statutory advisory body, the Animal Procedures Committee, to replace older arrangements.[50]

The administrative mechanisms designed by the 1876 Act provided the backbone of the late twentieth-century revised regulation. The 1986 Act only slightly amended the licensing requirements. Three licenses were now required: a personal license for the individual investigator, a license for the establishment in which the research would take place, and a project license detailing the number and type of animals to be used as well as the exact procedure to be performed and the overall purpose of the project.[51] Breeders and suppliers of certain species had to be registered also.[52]

Shortly before the enactment of the 1986 Act, Patrick Bateson, anticipating the new legislation, published "When to Experiment on Animals" in the *New Scientist*.[53] Bateson contended that the conflicts surrounding vivisection could be reconciled in "practical ways." He compared research on animals to shopping for a new pair of shoes: "you will want good quality and you will also wish to pay less. You will probably set an upper limit on how much you will pay and a lower limit on the quality, but the limit for one will depend on the other." He reasoned that similarly, "a much lower amount of suffering would be tolerated if the work were not regarded as being important" and vice versa. Bateson, therefore, presented the "decision cube," a model for harm/benefit analysis aiming to simplify the evaluation of experiments through the visual quantification of three criteria: quality of the research, certainty of medical benefit, and animal suffering.[54] Bateson's cube would later gain a central role in the European Union's conceptualization of its animal welfare scheme.

The Consolidated 1986 Act: Enter the Cephalopod

At the beginning of the twenty-first century, identifying pain was no longer just a challenge for Home Office civil servants trying to make sense of the law to address lawmakers' concerns. In July 2002, a Select Committee on Animals in

---

50. Animals (Scientific Procedures) Act, Sections 19–20. The advisory body established after the second Royal Commission was reconstituted in 1980 and renamed the Advisory Committee on Animal Experiments. The Statutory body was established on April 1, 1987.

51. Animals (Scientific Procedures) Act, Sections 3–8.

52. Animals (Scientific Procedures) Act, Section 7.

53. Patrick Bateson, "When to Experiment on Animals," *New Scientist* 109, no. 1496 (February 20, 1986): 30–32.

54. Bateson, "When to Experiment on Animals," 31, 32.

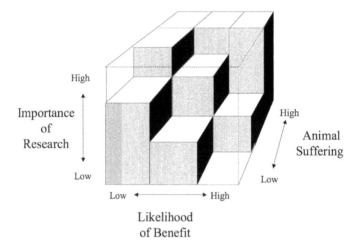

High

Importance
of
Research

Low

High

Animal
Suffering

Low

Low ◄———————► High

Likelihood
of Benefit

FIGURE 10 / Bateson's decision cube as a model for harm/benefit analysis in animal experimentation. From Pandora Pound and Christine J. Nicol, "Retrospective Harm Benefit Analysis of Pre-Clinical Animal Research for Six Treatment Interventions," PLOS ONE (2018), https://doi.org/10.1371/journal.pone.0193758.

Scientific Procedures called, among others, for the inspectorate to make serious efforts "to provide better statistics on animal suffering," adding that the inspectorate "should develop or approve a 'scoring system' for animal suffering."[55] The century-old efforts to define what counts for an experiment also remained. The Select Committee echoed concerns made by some witnesses that the current Act was arbitrary and "not always appropriate for experimental farm animals: taking a blood sample is a routine part of farm animal husbandry, but is counted as a procedure under the Act if carried out for experimental farm purposes."[56]

The recommendations of the Select Committee were incorporated into the Act only after the publication of the European Directive 2010/63/EU on the protection of animals used for scientific purposes, which came into force on November 9, 2010, replacing Directive 86/609/EEC. According to the European Directive, the changing science of sentience made it necessary to revisit legal standards: "New scientific knowledge is available in respect of factors influencing animal welfare as well as the capacity of animals to sense and express pain, suffering, distress and lasting harm. It is therefore necessary to improve the welfare of

55. Select Committee on Animals in Scientific Procedures, Volume 1—Report, July 16, 2002, HL 150-I 2001-02.

56. Select Committee on Animals in Scientific Procedures, Section 3.15.

animals used in scientific procedures by raising the minimum standards for their protection in line with the latest scientific developments."[57]

In 2012, the 1986 Act was amended to comply with the Directive.[58] According to the Home Office, many of the new Directive's provisions were "similar to current United Kingdom legislation and practice," and yet some changes had to be made in the British law.[59] The original 1986 Act defined protected animals as "any living vertebrate other than man," and a 1993 amended version included the common octopus (*Octopus vulgaris*) "from the stage of its development when it becomes capable of independent feeding."[60] The revised 2012 version covered "any living vertebrate other than man and any living cephalopod."[61] It followed the European Directive, which explained that "cephalopods should also be included in the scope of this Directive, as there is scientific evidence of their ability to experience pain, suffering, distress and lasting harm."[62]

The 2012 amended act also updated the definition of a "regulated procedure." In the revised definition, regulated procedures were those "which may have the effect of causing the animal a level of pain, suffering, distress or lasting harm equivalent to, or higher than, that caused by the introduction of a needle in accordance with good veterinary practice."[63] The needle test, which was first discussed during the vaccination debates (see chapter 3) resonated with the 2010 Directive, which did not apply to "practices not likely to cause pain, suffering, distress or lasting harm equivalent to, or higher than, that caused by the introduction of a needle in accordance with good veterinary practice."[64]

---

57. Directive 2010/63/EU of the European Parliament and of the Council of September 22, 2010 on the Protection of Animals Used for Scientific Purposes (referred to as European Directive 2010/63/EU), Section 6.

58. European Directive 2010/63/EU, Section 2.

59. Explanatory Memorandum to the Animals (Scientific Procedures) Act 1986 Amendment Regulations 2012, No. 3039, December 18, 2012, Draft, Subsection 7(7.3), http://www.legislation.gov.uk/ukdsi/2012/9780111530313/pdfs/ukdsiem_9780111530313_en.pdf.

60. Animals (Scientific Procedures) Act (Amendment) Order 1993.

61. Consolidated version of the Animals (Scientific Procedures) Act, 1986, c. 14, Section 1, https://www.gov.uk/government/uploads/system/uploads/attachment_data/file/308593/ConsolidatedASPA1Jan2013.pdf.

62. European Directive 2010/63/EU, Section 8.

63. The Animals (Scientific Procedures) Act 1986 Amendment Regulations 2012, http://www.legislation.gov.uk/ukdsi/2012/9780111530313/pdfs/ukdsi_9780111530313_en.pdf. According to Subsection 2(1)(1A) "A qualifying purpose" was applied for an experimental or another scientific purpose (whether or not the outcome of the procedure is known); or (b) it is applied for an educational purpose.

64. European Directive 2010/63/EU, Chapter 1, Article 1, Section 4(f).

Scales of Pain

A remarkable addition to the Animals (Scientific Procedures) Act 1986 was a list of factors that the home secretary was required to take into account when examining a license application. The 2012 amendment instructed the home secretary, among others, to evaluate the objectives of experiments and their predicted scientific benefits or educational value, and to assess their compliance with the 3Rs principle.[65] Furthermore, the home secretary was required to classify the likely severity of each regulated procedure as "non-recovery," "mild," "moderate," or "severe." The act instructed the home secretary to "carry out a harm-benefit analysis of the programme of work to assess whether the harm that would be caused to protected animals in terms of suffering, pain and distress is justified by the expected outcome, taking into account ethical considerations and the expected benefit to human beings, animals or the environment."[66]

A 2014 Guidance on the Operation of the Animal (Scientific Procedures) Act 1986 shed some light on the Home Office's way of assessing the severity of proposed experiments:

> We look at the types of procedure you are going to use considering particularly: the type of manipulation and handling; the nature of the pain, suffering, distress or lasting harm likely to be caused by the procedure; its intensity, duration, frequency and the number of techniques being used in each animal; any cumulative suffering within a protocol; if the animal is prevented from behaving naturally by restricting its housing, husbandry and standards of care; methods used to reduce or eliminate pain, suffering and distress, including refining housing, husbandry and care; and humane end-points and how they will be applied.

Furthermore, the guidance specifies "type of species, strain and genotype; the maturity, age and gender of the animal; training experience of the animal with respect to the procedure; and if the animal is to be re-used, the actual severity of the previous procedures" as elements in determining the severity of the procedure.[67]

The Home Office defined the categories of "mild," "moderate," "severe," and "non-recovery" pain as the following: under the "mild" category were

---

65. Amendment Regulations 2012, Subsections 5B(3)(a)–(b).
66. Amendment Regulations 2012, Subsections 5B(3)(c)–(d).
67. Home Office, Guidance on the Operation of Animal (Scientific Procedures) Act 1986, March 2014, unnumbered command paper, Subsection 5.12.2, https://www.gov.uk/government/uploads/system/uploads/attachment_data/file/291350/Guidance_on_the_Operation_of_ASPA.pdf.

"procedures on animals as the result of which the animals are likely to experience short-term mild pain, suffering or distress, as well as procedures with no significant impairment of the well-being or general condition of the animals." For example, noninvasive imaging of animals with appropriate sedation or anesthesia, or "breeding of genetically altered animals, which is expected to result in a phenotype with mild effects."[68]

In "moderate" procedures "the animals are likely to experience short-term moderate pain, suffering or distress, or long-lasting mild pain, suffering or distress as well as procedures that are likely to cause moderate impairment of the well-being or general condition of the animals."[69] Among the examples given were the "use of metabolic cage involving moderate restriction of movement over a prolonged period (up to five days,)" and the "creation of genetically altered animals through surgical procedure."[70]

"Severe" procedures are those in which "the animals are likely to experience severe pain, suffering or distress, or long-lasing moderate pain, suffering or distress as well as procedures that are likely to cause severe impairment of the well-being or general condition of the animals."[71] For example, "inescapable electric shocks" and "irradiation or chemotherapy with a lethal dose without reconstitution of immune system."[72] Under "non-recovery" procedures were classified procedures "which are performed entirely under general anaesthesia from which the animal shall not recover consciousness."[73]

The Reconfigurations of Pain and Ethics

The relationship between humans and other animals in the context of regulation, in the words of the 2002 Select Committee, "has evolved under the influence of modern technological and scientific developments and our growing

---

68. Home Office, Guidance on the Operation of Animal (Scientific Procedures) Act 1986, Appendix G, Section III.

69. Home Office, Guidance on the Operation of Animal (Scientific Procedures) Act 1986, Section I.

70. Home Office, Guidance on the Operation of Animal (Scientific Procedures) Act 1986, Section III.

71. Home Office, Guidance on the Operation of Animal (Scientific Procedures) Act 1986, Section I.

72. Home Office, Guidance on the Operation of Animal (Scientific Procedures) Act 1986, Section III.

73. Home Office, Guidance on the Operation of Animal (Scientific Procedures) Act 1986, Section I.

understandings of animal cognition and suffering."[74] This insight could not have been articulated without, to borrow historian Anita Guerrini's words, the "reconfiguration of the human and animal body" in terms of pain, which had triggered in the nineteenth-century a reconfiguration of the ethical relations with animals as well as of the making of knowledge.[75]

The Vivisection Act developed a language of animal ethics focusing on pain and its control through combined legal and scientific means. Recognizing pain was inherent to its management. At times, the Home Office and physiologists were at odds when reflecting on the issue. Their correspondence framed the shared terms in which the understanding of pain had evolved. Indeed, a 2007 edition of a book surveying British laboratory law for practitioners advised its readers: "Good communication with the inspector is crucial."[76]

In the century and a half since the first animal experimentation law was enacted, pain management had become an integral part of animal husbandry in laboratories, and in legal documents detailing pain scoring systems.[77] The many efforts to imagine and understand animal sentience provided local and temporal responses to problems of pain, while repeatedly failing to provide a conclusive and encompassing legal paradigm to annul it. In the archives that chronicle these efforts, substantial epistemological and moral dilemmas were generated by mundane attempts to match standard certificates to various operations. Law and science cocreated the bureaucracy of empathy: an attempt to systemize the understanding of animal suffering through administrative mechanisms. Examining representative dilemmas in the previous chapters detailed the shifting relations between science, morals, and legal order. Out of these crossings emerged the laboratory animal: produced, supplied, documented, anesthetized, and certified from life's beginning to its end.

---

74. Report of the Select Committee on Animals in Scientific Procedures, 5.

75. Guerrini, *The Courtiers' Anatomists*, 3.

76. Kevin Dolan, *Laboratory Animal Law: Legal Control of the Use of Animals in Research*, 2nd ed. (Oxford: Wiley-Blackwell, 2007), 42.

77. For example, see S. James Gaynor and William W. Muir, *Handbook of Veterinary Pain Management*, 2nd ed. (St. Louis, MO: Mosby, 2008).

POSTSCRIPT / "Can They Suffer?"

For two centuries, the Benthamite question was the basis for moral obligations toward nonhuman animals. Utilitarian reasoning, balancing animal pain against human gain, undergirded the regulation of animal experimentation from its outset through the twentieth century. The European Union's 2002 "decision cube," a model for cost/benefit assessment of painful procedures, epitomizes this reasoning.[1] For a society that perceived pain (rather than, for example, will, mind, or consciousness) as the only shared element between humans and animals that was worthy of consideration, eliminating pain was the ultimate means to solve moral questions raised by using animals. However, pain gradually lost its primacy as an exclusive moral criterion in animal ethics. By the turn of the twenty-first century, pain was joined by claims about animals' intrinsic values or special capacities.[2] The novelist John Maxwell Coetzee represented these alternatives through the protagonist of *The Lives of Animals* (1999), an acclaimed author named Elizabeth Costello. In a passionate speech about animal rights, Costello reveals to her listeners the horrors of animal exploitation while advocating for sympathy for their suffering through poetic imagination. The philosopher Peter Singer, who in 1975 articulated the utilitarian argument in favor of animal welfare in the influential *Animal Liberation*, offered a utilitarian critique of Coetzee's version of animal ethics. He employed, ironically, a fictional figure who speaks against the fogginess of the poetic argument and stresses: "You haven't given me any reason why painless killing would be wrong, if other animals take their place and lead an equally good life."[3]

The philosopher Martha Nussbaum picked up Singer's challenge, and in "The Moral Status of Animals" she explained that one of the major problems with

---

1. "Modified Bateson Cube," National Competent Authorities for the Implementation of Directive 2010/63/EU on the Protection of Animals Used for Scientific Purposes, Working Document on Project Evaluation and Brussels, September 18–19, 2013.

2. Tom Regan, *The Case for Animal Rights* (Berkeley: University of California Press, 1983).

3. Peter Singer, "Reflections," in *The Lives of Animals*, by J. M. Coetzee (Princeton: Princeton University Press, 2001), 85–92, 90.

utilitarian reasoning is that it aggregates together "diverse aspects of lives, reducing them all to experienced pain and pleasure."[4] Nussbaum offered to replace the reductive utilitarian pain calculus by considering animals' needs and abilities. These may include—alongside the avoidance of pain—movement, affection, health, community, dignity, and bodily integrity. Perhaps in a less structured way, but as forcefully, the historian and social theorist Donna Haraway revisited the Benthamite emphasis on suffering in *When Species Meet* (2007). Haraway asked, "how much more promise is in the questions, Can animals play? Or work? And even can I learn to play with *this* cat?"[5]

It is too early to know how transformative the attempts are to go beyond pain in animal ethics and how—if at all—they will be translated into law. Advancing a legal vision of these questions could broaden the scope of vivisection regulation to include standards beyond the reduction of pain. For example, they could add doses of affection, or social enrichments of the living environment in the lab. Another potential consequence is a shift in the decisions which species to protect, or in priorities among them, according to the levels of their cognitive capacities. Some of these changes are already being seen regarding great apes. In 1996, for example, the UK prohibited the experimental use of nonhuman primates.[6] In 2010, the EU banned experiments on great apes, "the closest species to human beings with the most advanced social and behavioral skills."[7] It remains legal to experiment on any other species.

Alternative approaches to traditional utilitarian animal ethics emerged against the backdrop of decades of scientific research into animal cognition and social behavior. Ethicists and lawyers are now harnessing the fruits of these studies to make new claims for the protection of nonhuman animals. Following the lessons of the history of the Vivisection Act, we can expect the coproduction of knowledge and normative order: the changes in moral thought will likely coincide with a transformation in scientific practice, legal language, and more

---

4. Martha C. Nussbaum, "The Moral Status of Animals," *Chronicle of Higher Education* 52, no. 22 (February 3, 2006): B6–8. For Nussbaum's capabilities approach, see Martha C. Nussbaum, *Frontiers of Justice: Disability, Nationality, Species Membership* (Cambridge, MA: Belknap Press of Harvard University Press, 2007).

5. Donna J. Haraway, *When Species Meet* (Minneapolis: University of Minnesota Press, 2007), 22; emphasis added. See also Cary Wolfe, "Flesh and Finitude: Thinking Animals in (Post) Humanist Philosophy," *SubStance* 37, no. 3 (2008): 8–36, http://www.jstor.org/stable/25195184.

6. House of Lords Select Committee, 2002, 10.

7. Directive 2010/63/EU of the European Parliament and of the Council of 22 September 2010 on the Protection of Animals Used for Scientific Purposes.

generally in the definition of expertise. What will replace the bureaucracy of empathy in a legal system in which dignity, community, or playfulness are as significant as pain? How we value animals, how we understand them, and how we regulate our relationships with them—these questions have been, are, and will remain intertwined.

BRITISH CRUELTY TO ANIMALS LEGISLATION AND CASE LAW, NINETEENTH CENTURY

CASE LAW	LEGISLATION
	1822 An Act to Prevent the Cruel and Improper Treatment of Cattle
	1833 An Act for the More Effectual Administration of Justice in the Office of a Justice of the Peace in the Several Police Offices established in the Metropolis, and for the More Effectual Prevention of Depredation on the River Thames and its Vicinity, for Three Years
	1835 An Act to Consolidate and Amend the Several Laws Relating to the Cruel and Improper Treatment of Animals, and the Mischiefs Arising from the Driving of Cattle, and to Make Other Provisions in Regard Thereto
	1839 An Act for Further Improving the Police in and Near the Metropolis
	1844 An Act to Amend the Law for Regulating Places Kept for Slaughtering Horses
	1848 An Act to Prohibit the Importation of Sheep, Cattle, or Other Animals for the Purpose of Preventing the Introduction of Contagious or Infectious Disorders
	1849 An Act for the More Effectual Prevention of Cruelty to Animals

(Continued)

(Continued)

CASE LAW	LEGISLATION
	1854 An Act to Amend an Act of the Twelfth and Thirteenth Years of Her Present Majesty for the More Effectual Prevention of Cruelty to Animals
1860 Clark v. Hague, (1860) 2 L.T. 85 (Q.B.)	
1863 Budge v. Parsons (1863) 3 B & S 382 (Q.B.) Morley v. Greenhalgh, (1863) 7 L.T. 624 (Q.B.)	
	1869 The Contagious Diseases (Animals) Act Sea Birds Preservation Act
1871 Colam v. Hall (1871) 6 Q.B. 206	
1874 Pitts v. Millar, (1874) 30 L.T. 328 (Q.B.)	
	1876 The Cruelty to Animals Act
1877 Murphy v. Manning, (1877) 2 Exch. D. 307.	
1878 Powell v. Knight, (1878) 38 L.T. 607 (Q.B.) Everitt v. Davies, (1878) 38 L.T. 360 (Ex. Div.)	
1881 Swan v. Sanders, (1881) 50 L.J. 67 (Q.B)	
1883 Colam v. Pagett, (1883) 12 Q.B.D. 66.	
1884 Brady v. McArdle (1884) 14 L.R.N. 174 (Exchequer Div.) (Ir.)	
1887 Lewis v. Fermor (1887) 18 Q.B.D. 532.	
1889 Ford v. Wiley, (1899) 23 Q.B.D. 203.	
1890 Adcock v. Murrell, (1890) 54 J.P. 776. Filburn v. The People Palace and Aquarium Co., (1890) 25 Q.B.D. 258.	
1893 Aplin v. Porritt, (1893) 2 Q.B.D. 57.	
1894 Harper v. Marcks, (1894) 2 Q. B. 319.	
1896 Yates v. Higgins, (1896) 65 L.J.M.C. 31 (Q.B.)	
1899 Duncan v. Pope, (1899) 80 L.T. 120 (Q.B.)	**-1900** Wild Animals in Captivity Protection Act

AN ACT TO AMEND THE LAW RELATING TO CRUELTY TO ANIMALS, 1876

Cruelty to Animals.

[39 & 40 Vict. Ch. **77.**]

ARRANGEMENT OF CLAUSES.

[*Public.-77.*] A

CHAPTER 77.

An Act to amend the Law relating to Cruelty to Animals. A.D. 1876.

[15th August 1876.]

WHEREAS it is expedient to amend the law relating to cruelty to animals by extending it to the cases of animals which for medical, physiological, or other scientific purposes are subjected when alive to experiments calculated to inflict pain :

Be it enacted by the Queen's most Excellent Majesty, by and with the advice and consent of the Lords Spiritual and Temporal, and Commons, in this present Parliament assembled, and by the authority of the same, as follows :

1. This Act may be cited for all purposes as "The Cruelty to Animals Act, 1876." Short title.

2. A person shall not perform on a living animal any experiment calculated to give pain, except subject to the restrictions imposed by this Act. Any person performing or taking part in performing any experiment calculated to give pain, in contravention of this Act, shall be guilty of an offence against this Act, and shall, if it be the first offence, be liable to a penalty not exceeding fifty pounds, and if it be the second or any subsequent offence, be liable, at the discretion of the court by which he is tried, to a penalty not exceeding one hundred pounds or to imprisonment for a period not exceeding three months. Prohibition of painful experiments on animals.

3. The following restrictions are imposed by this Act with respect to the performance on any living animal of an experiment calculated to give pain ; that is to say, General restrictions as to performance of painful experiments on animals.

    (1.) The experiment must be performed with a view to the advancement by new discovery of physiological knowledge or of knowledge which will be useful for saving or prolonging life or alleviating suffering ; and

    (2.) The experiment must be performed by a person holding such license from one of Her Majesty's Principal Secre-

A.D. 1876.

taries of State, in this Act referred to as the Secretary of State, as is in this Act mentioned, and in the case of a person holding such conditional license as is hereinafter mentioned, or of experiments performed for the purpose of instruction in a registered place ; and

(3.) The animal must during the whole of the experiment be under the influence of some anæsthetic of sufficient power to prevent the animal feeling pain ; and

(4.) The animal must, if the pain is likely to continue after the effect of the anæsthetic has ceased, or if any serious injury has been inflicted on the animal, be killed before it recovers from the influence of the anæsthetic which has been administered ; and

(5.) The experiment shall not be performed as an illustration of lectures in medical schools, hospitals, colleges, or elsewhere ; and

(6.) The experiment shall not be performed for the purpose of attaining manual skill.

Provided as follows ; that is to say,

(1.) Experiments may be performed under the foregoing provisions as to the use of anæsthetics by a person giving illustrations of lectures in medical schools, hospitals, or colleges, or elsewhere, on such certificate being given as in this Act mentioned, that the proposed experiments are absolutely necessary for the due instruction of the persons to whom such lectures are given with a view to their acquiring physiological knowledge or knowledge which will be useful to them for saving or prolonging life or alleviating suffering ; and

(2.) Experiments may be performed without anæsthetics on such certificate being given as in this Act mentioned that insensibility cannot be produced without necessarily frustrating the object of such experiments ; and

(3.) Experiments may be performed without the person who performed such experiments being under an obligation to cause the animal on which any such experiment is performed to be killed before it recovers from the influence of the anæsthetic on such certificate being given as in this Act mentioned, that the so killing the animal would necessarily frustrate the object of the experiment, and provided that the animal be killed as soon as such object has been attained ; and

2

(4.) Experiments may be performed not directly for the advancement by new discovery of physiological knowledge, or of knowledge which will be useful for saving or prolonging life or alleviating suffering, but for the purpose of testing a particular former discovery alleged to have been made for the advancement of such knowledge as last aforesaid, on such certificate being given as is in this Act mentioned that such testing is absolutely necessary for the effectual advancement of such knowledge.

4. The substance known as urari or curare shall not for the purposes of this Act be deemed to be an anæsthetic.

Use of urari as an anæsthetic prohibited.

5. Notwithstanding anything in this Act contained, an experiment calculated to give pain shall not be performed without anæsthetics on a dog or cat, except on such certificate being given as in this Act mentioned, stating, in addition to the statements herein-before required to be made in such certificate, that for reasons specified in the certificate the object of the experiment will be necessarily frustrated unless it is performed on an animal similar in constitution and habits to a cat or dog, and no other animal is available for such experiment; and an experiment calculated to give pain shall not be performed on any horse, ass, or mule except on such certificate being given as in this Act mentioned that the object of the experiment will be necessarily frustrated unless it is performed on a horse, ass, or mule, and that no other animal is available for such experiment.

Special restrictions on painful experiments on dogs, cats, &c.

6. Any exhibition to the general public, whether admitted on payment of money or gratuitously, of experiments on living animals calculated to give pain shall be illegal.

Absolute prohibition of public exhibition of painful experiments.

Any person performing or aiding in performing such experiments shall be deemed to be guilty of an offence against this Act, and shall, if it be the first offence, be liable to a penalty not exceeding fifty pounds, and if it be the second or any subsequent offence, be liable, at the discretion of the court by which he is tried, to a penalty not exceeding one hundred pounds or to imprisonment for a period not exceeding three months.

And any person publishing any notice of any such intended exhibition by advertisement in a newspaper, placard, or otherwise shall be liable to a penalty not exceeding one pound.

A person punished for an offence under this section shall not for the same offence be punishable under any other section of this Act.

3

A.D. 1876.

Administration of Law.

Registry of
place for
performance
of experi-
ments.

7. The Secretary of State may insert, as a condition of granting any license, a provision in such license that the place in which any experiment is to be performed by the licensee is to be registered in such manner as the Secretary of State may from time to time by any general or special order direct; provided that every place for the performance of experiments for the purpose of instruction under this Act shall be approved by the Secretary of State, and shall be registered in such manner as he may from time to time by any general or special order direct.

License by
Secretary of
State.

8. The Secretary of State may license any person whom he may think qualified to hold a license to perform experiments under this Act. A license granted by him may be for such time as he may think fit, and may be revoked by him on his being satisfied that such license ought to be revoked. There may be annexed to such license any conditions which the Secretary of State may think expedient for the purpose of better carrying into effect the objects of this Act, but not inconsistent with the provisions thereof.

Reports to
Secretary of
State.

9. The Secretary of State may direct any person performing experiments under this Act from time to time to make such reports to him of the result of such experiments, in such form and with such details as he may require.

Inspection
by Secretary
of State.

10. The Secretary of State shall cause all registered places to be from time to time visited by inspectors for the purpose of securing a compliance with the provisions of this Act, and the Secretary of State may, with the assent of the Treasury as to number, appoint any special inspectors, or may from time to time assign the duties of any such inspectors to such officers in the employment of the Government, who may be willing to accept the same, as he may think fit, either permanently or temporarily.

Certificate
of scientific
bodies for
exceptions to
general re-
gulations.

11. Any application for a license under this Act and a certificate given as in this Act mentioned must be signed by one or more of the following persons; that is to say,

    The President of the Royal Society;

    The President of the Royal Society of Edinburgh;

    The President of Royal Irish Academy;

    The Presidents of the Royal Colleges of Surgeons in London, Edinburgh, or Dublin;

    4

The Presidents of the Royal Colleges of Physicians in London,
Edinburgh, or Dublin;

The President of the General Medical Council;

The President of the Faculty of Physicians and Surgeons of
Glasgow;

The President of the Royal College of Veterinary Surgeons, or
the President of the Royal Veterinary College, London, but
in the case only of an experiment to be performed under
anæsthetics with a view to the advancement by new discovery
of veterinary science;

and also (unless the applicant be a professor of physiology, medicine, anatomy, medical jurisprudence, materia medica, or surgery in
a university in Great Britain or Ireland, or in University College,
London, or in a college in Great Britain or Ireland, incorporated
by royal charter) by a professor of physiology, medicine, anatomy,
medical jurisprudence, materia medica, or surgery in a university in
Great Britain or Ireland, or in University College, London, or in
a college in Great Britain or Ireland, incorporated by royal charter.

Provided that where any person applying for a certificate under
this Act is himself one of the persons authorised to sign such certificate, the signature of some other of such persons shall be
substituted for the signature of the applicant.

A certificate under this section may be given for such time
or for such series of experiments as the person or persons signing
the certificate may think expedient.

A copy of any certificate under this section shall be forwarded
by the applicant to the Secretary of State, but shall not be available
until one week after a copy has been so forwarded.

The Secretary of State may at any time disallow or suspend any
certificate given under this section.

12. The powers conferred by this Act of granting a license or giving **Power of**
a certificate for the performance of experiments on living animals judge to
grant license
may be exercised by an order in writing under the hand of any for experi-
judge of the High Court of Justice in England, of the High Court ment when
necessary in
of Session in Scotland, or of any of the superior courts in Ireland, criminal case.
including any court to which the jurisdiction of such last-mentioned
courts may be transferred, in a case where such judge is satisfied
that it is essential for the purposes of justice in a criminal case to
make any such experiment.

Legal Proceedings.

13. A justice of the peace, on information on oath that there is Entry on
reasonable ground to believe that experiments in contravention of warrant by
justice.

5

A.D. 1876.   this Act are being performed by an unlicensed person in any place not registered under this Act may issue his warrant authorising any officer or constable of police to enter and search such place, and to take the names and addresses of the persons found therein.

Any person who refuses admission on demand to a police officer or constable so authorised, or obstructs such officer or constable in the execution of his duty under this section, or who refuses on demand to disclose his name or address, or gives a false name or address, shall be liable to a penalty not exceeding five pounds.

Prosecution of offences and recovery of penalties in England.

14. In England, offences against this Act may be prosecuted and penalties under this Act recovered before a court of summary jurisdiction in manner directed by the Summary Jurisdiction Act.

In England " Summary Jurisdiction Act " means the Act of the session of the eleventh and twelfth years of the reign of Her present Majesty, chapter forty-three, intituled " An Act to " facilitate the performance of the duties of justices of the " peace out of sessions within England and Wales with respect " to summary convictions and orders," and any Act amending the same.

" Court of summary jurisdiction."

" Court of summary jurisdiction " means and includes any justice or justices of the peace, metropolitan police magistrate, stipendiary or other magistrate, or officer, by whatever name called, exercising jurisdiction in pursuance of the Summary Jurisdiction Act : Provided that the court when hearing and determining an information under this Act shall be constituted either of two or more justices of the peace in petty sessions, sitting at a place appointed for holding petty sessions, or of some magistrate or officer sitting alone or with others at some court or other place appointed for the administration of justice, and for the time being empowered by law to do alone any act authorised to be done by more than one justice of the peace.

Power of offender in England to elect to be tried on indictment, and not by summary jurisdiction.

15. In England, where a person is accused before a court of summary jurisdiction of any offence against this Act in respect of which a penalty of more than five pounds can be imposed, the accused may, on appearing before the court of summary jurisdiction, declare that he objects to being tried for such offence by a court of summary jurisdiction, and thereupon the court of summary jurisdiction may deal with the case in all respects as if the accused were charged with an indictable offence and not an offence punishable on summary conviction, and the offence may be prosecuted on indictment accordingly.

6

16. In England, if any party thinks himself aggrieved by any A.D. 1876. conviction made by a court of summary jurisdiction on determining any information under this Act, the party so aggrieved may appeal therefrom, subject to the conditions and regulations following :

(1.) The appeal shall be made to the next court of general or quarter sessions for the county or place in which the cause of appeal has arisen, holden not less than twenty-one days after the decision of the court from which the appeal is made ; and

(2.) The appellant shall, within ten days after the cause of appeal has arisen, give notice to the other party and to the court of summary jurisdiction of his intention to appeal, and of the ground thereof ; and

(3.) The appellant shall, within three days after such notice, enter into a recognizance before a justice of the peace, with two sufficient sureties, conditioned personally to try such appeal, and to abide the judgment of the court thereon, and to pay such costs as may be awarded by the court, or give such other security by deposit of money or otherwise as the justice may allow ; and

(4.) Where the appellant is in custody the justice may, if he think fit, on the appellant entering into such recognizance or giving such other security as aforesaid, release him from custody ; and

(5.) The court of appeal may adjourn the appeal, and upon the hearing thereof they may confirm, reverse, or modify the decision of the court of summary jurisdiction, or remit the matter to the court of summary jurisdiction with the opinion of the court of appeal thereon, or make such other order in the matter as the court thinks just, and if the matter be remitted to the court of summary jurisdiction the said last-mentioned court shall thereupon re-hear and decide the information in accordance with the order of the said court of appeal. The court of appeal may also make such order as to costs to be paid by either party as the court thinks just.

17. In Scotland, offences against this Act may be prosecuted and penalties under this Act recovered under the provisions of the Summary Procedure Act, 1864, or if a person accused of any offence against this Act in respect of which a penalty of more than five pounds can be imposed, on appearing before a court of summary jurisdiction, declare that he objects to being tried for such offence

A.D. 1876. in the court of summary jurisdiction, proceedings may be taken
against him on indictment in the Court of Justiciary in Edinburgh
or on circuit.

Every person found liable in any penalty or costs shall be
liable in default of immediate payment to imprisonment for a term
not exceeding three months, or until such penalty or costs **are**
sooner paid.

Prosecution of offences and recovery of penalties in Ireland.

18. In Ireland, offences against this Act may be prosecuted and
penalties under this Act recovered in a summary manner, subject
and according to the provisions with respect to the prosecution of
offences, the recovery of penalties, and to appeal of the Petty
Sessions (Ireland) Act, 1851, and any Act amending the same,
and in Dublin of the Acts regulating the powers of justices of the
peace or of the police of Dublin metropolis. All penalties recovered under this Act shall be applied in manner directed by the
Fines (Ireland) Act, 1871, and any Act amending the same.

Power of offender in Ireland to elect to be tried on in-dictment, and not by summary jurisdiction.

19. In Ireland, where a person is accused before a court of summary jurisdiction of any offence against this Act in respect of which
a penalty of more than five pounds can be imposed, the accused
may, on appearing before the court of summary jurisdiction,
declare that he objects to being tried for such offence by a court
of summary jurisdiction, and thereupon the court of summary
jurisdiction may deal with the case in all respects as if the accused
were charged with an indictable offence and not an offence
punishable on summary conviction, and the offence may be
prosecuted on indictment accordingly.

Interpreta-tion of " the Secretary of State " as to Ireland.

20. In the application of this Act to Ireland the term " the
Secretary of State " shall be construed to mean the Chief Secretary
to the Lord Lieutenant of Ireland for the time being.

Prosecution only with leave of Secretary of State.

21. A prosecution under this Act against a licensed person shall
not be instituted except with the assent in writing of the Secretary
of State.

Not to apply to in-vertebrate animals.

22. This Act shall not apply to invertebrate animals.

PRINTED IN ENGLAND FOR J A DOLE
Controller and Chief Executive of Her Majesty's Stationery Office and
Queen's Printer of Acts of Parliament
Reprinted in the Standard Parliamentary Page Size
1st Impression 1876
11th Impression January 1979

ISBN 0 10 850319 4

BIBLIOGRAPHY

Alberti, Fay Bound. *Medicine, Emotion and Disease, 1700–1950*. Houndmills: Palgrave Macmillan, 2006.

Ankeny, Rachel A., and Sabina Leonelli. "What's So Special about Model Organisms?" *Studies in History and Philosophy of Science Part A* 42, no. 2 (June 2011): 313–23.

"Application of Chloroform to Animals." *Illustrated London News*, December 11, 1847. The Illustrated London News Historical Archive, 1842–2003.

Asdal, Kristin. "Subjected to Parliament: The Laboratory of Experimental Medicine and the Animal Body." *Social Studies of Science* 38, no. 6 (2008): 899–917.

Aydede, Murat, and Güven Güzeldere. "Some Foundational Problems in the Scientific Study of Pain." *Philosophy of Science* 69, no. S3 (2002): S265–S83.

Bandes, Susan. "Empathy, Narrative, and Victim Impact Statements." *University of Chicago Law Review* 63, no. 2 (1996): 361–412.

Bargheer, Stefan. "The Fools of the Leisure Class: Honor, Ridicule, and the Emergence of Animal Protection Legislation in England, 1740–1840." *Archives Européennes de Sociologie* 47, no. 1 (2006): 3–35.

Bartrip, P. W. J. "State Intervention in Mid-Nineteenth Century Britain: Fact or Fiction?" *Journal of British Studies* 23, no. 1 (October 1, 1983): 63–83.

Bates, A. W. H. *Vivisection, Virtue, and the Law in the Nineteenth Century*. London: Palgrave Macmillan, 2017.

Bateson, Patrick. "When to Experiment on Animals." *New Scientist (1971)* 109, no. 1496 (February 20, 1986): 30–32.

Bekoff, Marc, and Carron A. Meaney, eds. *Encyclopedia of Animal Rights and Animal Welfare*. New York: Routledge, 2013.

Bentham, Jeremy. *An Introduction to the Principles of Morals and Legislation*. Hafner Publishing Company, (1789) 1948.

Benson, Etienne. "Animal Writes: Historiography, Disciplinarity, and the Animal Trace." In *Making Animal Meaning*, edited by Linda Kalof and Georgina M. Montgomery, 3–18. Animal Turn. East Lansing: Michigan State University Press, 2011.

Biagioli, Mario, ed. *The Science Studies Reader*. New York: Routledge, 1999.

Bittle, Carla. "Science, Suffrage, and Experimentation: Mary Putnam Jacobi and the Controversy over Vivisection in Late Nineteenth-Century America." *Bulletin of the History of Medicine* 79, no. 4 (2005): 664–94.

Boddice, Rob. *Science of Sympathy*. Urbana: University of Illinois Press, 2016.

——. "Species of Compassion: Aesthetics, Anaesthetics, and Pain in the Physiological Laboratory." *19: Interdisciplinary Studies in the Long Nineteenth Century*, no. 15 (June 12, 2012).

——. "The History of Emotions: Past, Present, Future." *Revista de Estudios Sociales* 62 (October 1, 2017): 10–15.

——. "Vivisecting Major: A Victorian Gentleman Scientist Defends Animal Experimentation, 1876–1885." *Isis* 102, no. 2 (June 1, 2011): 215–37.

Borell, Merriley. "Instrumentation and the Rise of Modern Physiology." *Science & Technology Studies* 5, no. 2 (1987): 53–62.

Bourke, Joanna. "Pain Sensitivity: An Unnatural History from 1800 to 1965." *Journal of Medical Humanities* 35, no. 3 (2014): 301–19.

——. *The Story of Pain: From Prayer to Painkillers*. New York: Oxford University Press, 2014.

Braverman, Irus, ed. *Animals, Biopolitics, Law: Lively Legalities*. London: Routledge, 2015.

Browne, Janet. "Forward." In *Medicine, Emotion and Disease, 1700–1950*, edited by Fay Bound Alberti, ix. New York: Palgrave Macmillan, 2006.

Browne, Janet. *Charles Darwin: A Biography. Vol. 2: The Power of Place*. Princeton: Princeton University Press, 2003.

Bucchi, Massimiano, and Federico Neresini. "Science and Public Participation." In *The Handbook of Science and Technology Studies*, edited by Edward J. Hackett, Olga Amsterdamska, Michael E. Lynch, and Judy Wajcman, 449–72. Cambridge: MIT Press, 2007.

Burnett, D. Graham. *Trying Leviathan: The Nineteenth-Century New York Court Case That Put the Whale on Trial and Challenged the Order of Nature*. Princeton: Princeton University Press, 2010.

Butler, Stella V. F. "Centers and Peripheries: The Development of British Physiology, 1870–1914." *Journal of the History of Biology* 21, no. 3 (October 1, 1988): 473–500.

Bynum, W. F. "'C'est Un Malade': Animal Models and Concepts of Human Diseases1." *Journal of the History of Medicine and Allied Sciences* 45, no. 3 (July 1, 1990): 397–413.

Capozzola, Christopher Joseph Nicodemus. *Uncle Sam Wants You: World War I and the Making of the Modern American Citizen*. New York: Oxford University Press, 2010.

Carpenter, Daniel. *The Forging of Bureaucratic Autonomy: Reputations, Networks, and Policy Innovation in Executive Agencies, 1862–1928*. Princeton: Princeton: University Press, 2001.

Cartwright, Lisa. "'Experiments of Destruction': Cinematic Inscriptions of Physiology." *Representations*, no. 40 (October 1, 1992): 129–52.

Cash, J. Theodore, and Wyndham R. Dunstan. "The Pharmacology of Aconitine, Diacetyl-Aconitine, Benzaconine, and Aconine, Considered in Relation to Their Chemical Constitution." *Philosophical Transactions of the Royal Society of London. Series B, Containing Papers of a Biological Character* 190 (January 1, 1898): 239–393.

Caton, Donald. *What a Blessing She Had Chloroform: The Medical and Social Response to the Pain of Childbirth from 1800 to the Present*. New Haven: Yale University Press, 1999.

Chakrabarti, Pratik. "Beasts of Burden: Animals and Laboratory Research in Colonial India." *History of Science* 48, no. 2 (June 1, 2010): 125–51.

Clarac, François. "Some Historical Reflections on the Neural Control of Locomotion." *Brain Research Reviews* 57, no. 1 (January 2008): 13–21.

Clark, Michael, and Catherine Crawford, eds. *Legal Medicine in History*. Cambridge: Cambridge University Press, 1994.

Clarke, Adele E. "Research Materials and Reproductive Science in the United States, 1910–1940." In *Physiology in the American Context 1850–1940*, edited by Gerald L. Geison, 323–50. New York: Springer, 1987.

Clarke, Edwin, and L. S. Jacyna. *Nineteenth-Century Origins of Neuroscientific Concepts*. Berkeley: University of California Press, 1987.

Cobbe, Frances Power. *Schadenfreude (Pleasure in Pain of Others): A Study*. London, 1902.

Cobbe, Frances Power, and Benjamin Bryan. *Vivisection in America: I. How It Is Taught II. How It Is Practiced*. London: Swan Sonnenschein, 1890.

Coetzee, J. M. *The Lives of Animals:* Princeton: Princeton University Press, 2001.

Coleman, William, and Frederic L. Holmes, eds. *The Investigative Enterprise: Experimental Physiology in Nineteenth-Century Medicine*. Berkeley: University of California Press, 1988.

Coleridge, John Duke Coleridge. *The Lord Chief Justice of England on Vivisection*. Nineteenth Century Collections Online: British Politics and Society. London: Office of the Society for Protection of Animals from Vivisection, 1882.

Coplan, Amy, and Peter Goldie, eds. *Empathy: Philosophical and Psychological Perspectives*. Oxford: Oxford University Press, 2014.

Cowper, Cecil, ed. "Notes and News." *Academy and Literature, 1910–1914*, no. 2137 (April 19, 1913): 504–5.

Cromwell, Valerie. "Interpretations of Nineteenth-Century Administration: An Analysis." *Victorian Studies* 9, no. 3 (March 1, 1966): 245–55.

Crook, Tom. "Sanitary Inspection and the Public Sphere in Late Victorian and Edwardian Britain: A Case Study in Liberal Governance." *Social History* 32, no. 4 (November 1, 2007): 369–93.

Cunningham, Andrew, and Perry Williams. *The Laboratory Revolution in Medicine*. Cambridge; Cambridge University Press, 2002.

Darwall, Stephen. "Empathy, Sympathy, Care." *Philosophical Studies: An International Journal for Philosophy in the Analytic Tradition* 89, no. 2/3 (1998): 261–82.

Daston, Lorraine J., and Peter Galison. *Objectivity*. New York: Zone Books, 2010.

Daston, Lorraine, and Gregg Mitman. *Thinking with Animals: New Perspectives on Anthropomorphism*. New York: Columbia University Press, 2005.

Daston, Lorraine, and Katharine Park. *Wonders and the Order of Nature, 1150–1750*. New York: Zone Books, 2001.

Desmond, Adrian J. *The Politics of Evolution: Morphology, Medicine, and Reform in Radical London*. Chicago: University of Chicago Press, 1989.

Dicey, A. V. *Lectures on the Relation between Law and Public Opinion in England*. Indianapolis, IN: Liberty Fund Inc., 2008.

Dolan, Kevin. *Laboratory Animal Law: Legal Control of the Use of Animals in Research*. Oxford: Wiley-Blackwell, 2007.

Dror, Otniel E. "Techniques of the Brain and the Paradox of Emotions, 1880–1930." *Science in Context* 14, no. 4 (December 2001): 643–60.

———. "The Affect of Experiment: The Turn to Emotions in Anglo-American Physiology, 1900–1940." *Isis* 90, no. 2 (June 1, 1999): 205–37.

Druglitrø, Tone. "'Skilled Care' and the Making of Good Science." *Science, Technology, and Human Values* 43, no. 4 (2018): 649–70.

Durbach, Nadja. *Bodily Matters: The Anti-Vaccination Movement in England, 1853–1907*. Durham: Duke University Press Books, 2004.

Elston, Mary Ann. "Women and Anti-Vivisection in Victorian England, 1870–1900." In *Vivisection in Historical Perspective*, edited by Nicolaas A Rupke, 259–94. London: Routledge, 1990.

Endersby, Jim. *A Guinea Pig's History of Biology: The Plants and Animals Who Taught Us the Facts of Life*. New York: Arrow Books, 2009.

——. "Sympathetic Science: Charles Darwin, Joseph Hooker, and the Passions of Victorian Naturalists." *Victorian Studies* 51, no. 2 (2009): 299–320.

Erichsen, John Eric. "The Antivivisectionists and the Progress of Modern Surgery." *British Medical Journal* 2, no. 985 (November 15, 1879): 794.

——. *The Member, the Fellow and the Franchise [in the Royal College of Surgeons]*. London: HKLewis, 1886.

Ferrier, David. "Experiments on the Brain of Monkeys.—No. I." *Proceedings of the Royal Society of London* 23 (1874): 409–30.

——. "The Croonian Lecture.—Experiments on the Brain of Monkeys (Second Series)." *Philosophical Transactions of the Royal Society of London* 165 (1875): 433–88.

——. *The Functions of the Brain*. New York: G.P. Putnam's Sons, 1876.

Ferrier, David, and Gerald F. Yeo. "A Record of Experiments on the Effects of Lesion of Different Regions of the Cerebral Hemispheres." *Philosophical Transactions of the Royal Society of London* 175 (1884): 479–564.

——. "The Functional Relations of the Motor Roots of the Brachial and Lumbo-Sacral Plexuses." *Proceedings of the Royal Society of London* 32 (1881): 12–20.

Finger, Stanley. *Minds behind the Brain: A History of the Pioneers and Their Discoveries*. Oxford: Oxford University Press, 2004.

Finn, Michael A., and James F. Stark. "Medical Science and the Cruelty to Animals Act 1876: A Re-Examination of Anti-Vivisectionism in Provincial Britain." *Studies in History and Philosophy of Science Part C: Studies in History and Philosophy of Biological and Biomedical Sciences* 49 (February 1, 2015): 12–23.

Flint, Austin. *Handbook of Physiology*. New York: Macmillan, 1905.

Franklin, R. J. M. "The Brown Animal Sanatory Institution—Historical Lessons for the Present?" *Veterinary Journal* 159, no. 3 (May 2000): 231–37.

Franklin, Sarah. *Dolly Mixtures: The Remaking of Genealogy*. Durham: Duke University Press, 2007.

French, Richard. *Antivivisection and Medical Science in Victorian Society*. Princeton: Princeton University Press, 1975.

Fudge, Erica. *Brutal Reasoning: Animals, Rationality, and Humanity in Early Modern England*. Ithaca, NY: Cornell University Press, 2006.

Fujimura, Joan H. "Standardizing Practices: A Socio-History of Experimental Systems in Classical Genetic and Virological Cancer Research, ca. 1920–1978." *History and Philosophy of the Life Sciences*, no. 1 (1996): 3–54.

Garner, Robert. *Animals, Politics and Morality*. Manchester: Manchester University Press, 2004.

Gaynor, S. James, and William W. Muir. *Handbook of Veterinary Pain Management*. 2nd ed. St. Louis, MO: Mosby, 2008.

Geison, Gerald L. "The Protoplasmic Theory of Life and the Vitalist-Mechanist Debate." *Isis* 60, no. 3 (1969): 273–92.

Gere, Cathy. *Pain, Pleasure, and the Greater Good: From the Panopticon to the Skinner Box and Beyond*. Chicago: University of Chicago Press, 2017.

Gieryn, Thomas F. *Cultural Boundaries of Science: Credibility on the Line*. Chicago: University of Chicago Press, 1999.

Gilbert, Pamela K. *Victorian Skin: Surface, Self, History*. Ithaca, NY: Cornell University Press, 2019.

Golan, Tal. *Laws of Men and Laws of Nature: The History of Scientific Expert Testimony in England and America*. Cambridge, MA: Harvard University Press, 2007.

Gooday, Graeme. "Placing or Replacing the Laboratory in the History of Science?" *Isis* 99, no. 4 (2008): 783–95.

Goodfield, G. J. *The Growth of Scientific Physiology: Physiological Method and the Mechanist-Vitalist Controversy, Illustrated by the Problems of Respiration and Animal Heat*. London: Hutchinson, 1960.

Gosling, J. C. B., and C. C. W. Taylor. *The Greeks on Pleasure*. Oxford: Oxford University Press, 1982.

Gray, Liz. "Body, Mind and Madness: Pain in Animals in Nineteenth-Century Comparative Psychology." In *Pain and Emotion in Modern History*, edited by Rob Boddice, 148–63. London: Palgrave Macmillan, 2014.

Great Britain, Parliament, House of Lords, and Select Committee on the Laboratory Animals Protection Bill (H.L.). *Report of the Select Committee on the Laboratory Animals Protection Bill (H.L.)*. Volume 1. London: H.M.S.O., 1980.

Greenfield, W. S. "An Inaugural Address on Pathology, Past and Present." *British Medical Journal* 2, no. 1088 (1881): 731–34.

Grier, Katherine C. *Pets in America: A History*. Orlando: Harvest Books, 2007.

Gross, Michael. "The Lessened Locus of Feelings: A Transformation in French Physiology in the Early Nineteenth Century." *Journal of the History of Biology* 12, no. 2 (1979): 231–71.

Gruen, Lori. *Entangled Empathy: An Alternative Ethic for Our Relationships with Animals*. New York: Lantern Books, 2015.

Guerrini, Anita. *Experimenting with Humans and Animals: From Galen to Animal Rights*. Baltimore: Johns Hopkins University Press, 2003.

——. *The Courtiers' Anatomists: Animals and Humans in Louis XIV's Paris*. Chicago: University of Chicago Press, 2015.

——. "The Ethics of Animal Experimentation in Seventeenth-Century England." *Journal of the History of Ideas* 50, no. 3 (1989): 391–407.

Hall, Marshall. *A Critical and Experimental Essay on the Circulation of the Blood*. London: Sherwood, Gilbert and Piper, 1831.

——. "Memoirs on Some Principles of Pathology in the Nervous System." *Medico-Chirurgical Transactions* 22 (1839): 191–217.

Halttunen, Karen. "Humanitarianism and the Pornography of Pain in Anglo-American Culture." *American Historical Review* 100, no. 2 (April 1, 1995): 303–34.

Hamilton, Susan. *Animal Welfare & Anti-Vivisection 1870–1910: Frances Power Cobbe*. New York: Routledge, 2004.

——. "Reading and the Popular Critique of Science in the Victorian Anti-Vivisection Press: Frances Power Cobbe's Writing for the Victoria Street Society." *Victorian Review* 36, no. 2 (October 1, 2010): 66–79.

——. "'Still Lives: Gender and the Literature of the Victorian Vivisection Controversy." *Victorian Review* 17, no. 2 (1991): 21–34.

Hamlin, Christopher. "Nuisances and Community in Mid-Victorian England: The Attractions of Inspection." *Social History* 38, no. 3 (2013): 346–79.

Haraway, Donna J. *Modest_Witness@Second_Millennium.FemaleMan_Meets_Onco-Mouse: Feminism and Technoscience*. New York: Routledge, 1997.

——. *Primate Visions: Gender, Race, and Nature in the World of Modern Science*. Reprint. Routledge, 1990.

——. *When Species Meet*. Minneapolis: University of Minnesota Press, 2007.

Harrison, Brian. "Animals and the State in Nineteenth-Century England." *English Historical Review* 88, no. 349 (October 1, 1973): 786–820.

Harvey, Caitlin. "Science and Sensibility: Louise Lind-Af-Hageby's Diary as Female Testimony, Scientific Publication, and Antivivisectionist Tool, 1890–1918." *Journal of Women's History* 30, no. 1 (2018): 80–106.

Helmreich, Stefan. *Alien Ocean: Anthropological Voyages in Microbial Seas*. University of California Press, 2009.

Helmreich, Stefan, and Sophia Roosth. "Life Forms: A Keyword Entry." *Representations* 112, no. 1 (November 1, 2010): 27–53.

Hennock, E. P. "Vaccination Policy Against Smallpox, 1835–1914: A Comparison of England with Prussia and Imperial Germany." *Social History of Medicine* 11, no. 1 (April 1, 1998): 49–71.

Herzig, Rebecca M. *Suffering for Science: Reason and Sacrifice in Modern America*. New Brunswick, NJ: Rutgers University Press, 2005.

Hobbins, Peter. *Venomous Encounters: Snakes, Vivisection and Scientific Medicine in Colonial Australia*. Manchester: Manchester University Press, 2017.

Hodder, Edwin. *The Life and Work of the 7th Earl of Shaftesbury, K.G.* Cassell & Company, 1887.

Hoff, H. E., and L. A. Geddes. "Graphic Registration before Ludwig; The Antecedents of the Kymograph." *Isis* 50, no. 1 (March 1, 1959): 5–21.

Holmes, Frederic L. "The Old Martyr of Science: The Frog in Experimental Physiology." *Journal of the History of Biology* 26, no. 2 (July 1, 1993): 311–28.

Horkheimer, Max, and Theodor W. Adorno. *Dialectic of Enlightenment: Philosophical Fragments*. Edited by Gunzelin Schmid Noerr. Translated by Edmund Jephcott. Cultural Memory in the Present. Stanford, CA: Stanford University Press, 2002.

Hume, L. J. "Jeremy Bentham and the Nineteenth-Century Revolution in Government." *Historical Journal* 10, no. 3 (1967): 361–75.

Hurren, Elizabeth T. *Dying for Victorian Medicine: English Anatomy and Its Trade in the Dead Poor, c.1834–1929*. Basingstoke: Palgrave Macmillan, 2014.

Hutton, Richard Holt. "The Biologists on Vivisection." *Nineteenth Century* 11 (1882): 29–39.

Jacobs, Noortje. "A Moral Obligation to Proper Experimentation: Research Ethics as Epistemic Filter in the Aftermath of World War II." *Isis* 111, no. 4 (December 2, 2020): 759–80.

Jasanoff, Sheila. "Beyond Epistemology: Relativism and Engagement in the Politics of Science." *Social Studies of Science* 26, no. 2 (1996): 393–418.

——. *The Fifth Branch: Science Advisers as Policymakers*. Cambridge, MA: Harvard University Press, 1990.

——. "Making Order: Law and Science in Action." In *The Handbook of Science and Technology Studies*, edited by Edward J. Hackett, Olga Amsterdamska, Michael

E. Lynch, Judy Wajcman, and Wiebe E. Bijker, 761–86. 3rd ed. Cambrdige: MIT Press, 2007.

——. *Science at the Bar: Law, Science and Technology in American Law*. Cambridge, MA: Harvard University Press, 1997.

Jasanoff, Sheila and Society for Social Studies of Science. *Handbook of Science and Technology Studies*. Thousand Oaks, CA: Sage, 1995.

Jones, Susan D. *Valuing Animals: Veterinarians and Their Patients in Modern America*. Annotated edition. Baltimore: Johns Hopkins University Press, 2002.

Kalof, Linda. *Looking at Animals in Human History*. London: Reaktion Books, 2007.

Kean, Hilda. *Animal Rights: Political and Social Change in Britain since 1800*. London: Reaktion Books, 1998.

——. "The 'Smooth Cool Men of Science': The Feminist and Socialist Response to Vivisection." *History Workshop Journal*, no. 40 (October 1, 1995): 16–38.

Kelly, Catherine, and Imogen Goold, eds. *Lawyers' Medicine: The Legislature, the Courts and Medical Practice, 1760–2000*. Oxford: Hart, 2009.

Kete, Kathleen. *The Beast in the Boudoir: Petkeeping in Nineteenth-Century Paris*. Berkeley: University of California Press, 1995.

Kimmelman, Barbara A. "Organisms and Interests in Scientific Research: R. A. Emerson's Claims for the Unique Contributions of Agricultural Genetics." In *The Right Tools for the Job*, edited by Adele E. Clarke and Joan H. Fujimura. Princeton, NJ: Princeton University Press, 1992.

Kirk, Robert G. W. "A Brave New Animal for a Brave New World: The British Laboratory Animals Bureau and the Constitution of International Standards of Laboratory Animal Production and Use, circa 1947–1968." *Isis: An International Review Devoted to the History of Science and Its Cultural Influences* 101, no. 1 (March 2010): 62–94.

——. "The Invention of the 'Stressed Animal' and the Development of a Science of Animal Welfare, 1947–86." In *Stress, Shock, and Adaptation in the Twentieth Century*, edited by David Cantor and Edmund Ramsden, 241–63. Rochester, NY: University of Rochester Press, 2014.

Kirk, Robert G. W., and Michael Worboys. "Medicine and Species: One Medicine, One History?" In *The Oxford Handbook of the History of Medicine*, edited by Mark Jackson, 561–77. Oxford: Oxford University Press, 2011.

Kirk, Robert G. W., and Neil Pemberton. "Re-Imagining Bleeders: The Medical Leech in the Nineteenth Century Bloodletting Encounter." *Medical History* 55, no. 3 (July 2011): 355–60.

Klein, Edward, and John Burdon-Sanderson. *Handbook for the Physiological Laboratory*. Philadelphia: Lindsay & Blakiston, 1873.

Kohler, Robert E. *Lords of the Fly: Drosophila Genetics and the Experimental Life*. Chicago: University of Chicago Press, 1994.

Landry, Donna. *Noble Brutes: How Eastern Horses Transformed English Culture*. Baltimore: Johns Hopkins University Press, 2008.

Lansbury, Coral. *The Old Brown Dog: Women, Workers, and Vivisection in Edwardian England*. Madison: University of Wisconsin Press, 1985.

Latour, Bruno, and Steve Woolgar. *Laboratory Life: The Construction of Scientific Facts*. Princeton: Princeton University Press, 1986.

Lavi, Shai J. *The Modern Art of Dying: A History of Euthanasia in the United States.* Princeton, NJ: Princeton University Press, 2005.

Lederer, Susan E. "Political Animals: The Shaping of Biomedical Research Literature in Twentieth-Century America." *Isis* 83, no. 1 (March 1, 1992): 61–79.

——. *Subjected to Science: Human Experimentation in America before the Second World War.* Baltimore: Johns Hopkins University Press, 1997.

Leigh Star, Susan. *Regions of the Mind: Brain Research and the Quest for Scientific Certainty.* Stanford, CA: Stanford University Press, 1989.

——. "Scientific Work and Uncertainty." *Social Studies of Science* 15, no. 3 (1985): 391–427.

Leneman, Leah. "The Awakened Instinct: Vegetarianism and the Women's Suffrage Movement in Britain." *Women's History Review* 6, no. 2 (June 1, 1997): 271–87.

Lind-af-Hageby, Lizzy, and Leisa Katherina Schartau. *The Shambles of Science: Extracts from the Diary of Two Students of Physiology.* London: E. Bell, 1903.

Littlewood, Sydney, United Kingdom, Home Office, and Departmental Committee on Experiments on Animals. "Report of the Departmental Committee on Experiments on Animals." London: Her Majesty's Stationery Office, 1965.

Lock, Margaret. *Twice Dead: Organ Transplants and the Reinvention of Death.* Berkeley: University of California Press, 2002.

Logan, Cheryl A. "'[A]Re Norway Rats . . . Things?': Diversity versus Generality in the Use of Albino Rats in Experiments on Development and Sexuality." *Journal of the History of Biology* 34, no. 2 (July 1, 2001): 287–314.

——. "Before There Were Standards: The Role of Test Animals in the Production of Empirical Generality in Physiology." *Journal of the History of Biology* 35, no. 2 (July 1, 2002): 329–63.

London Anti-Vivisection Society. *Vivisection. Analysis of the Report of the Royal Commission with Observations Thereon, and Extracts from the Evidence.* London, 1877.

Lynch, Michael E. "Sacrifice and the Transformation of the Animal Body into a Scientific Object: Laboratory Culture and Ritual Practice in the Neurosciences." *Social Studies of Science* 18, no. 2 (May 1, 1988): 265–89.

MacDonagh, Oliver. "The Nineteenth-Century Revolution in Government: A Reappraisal." *Historical Journal* 1, no. 1 (January 1, 1958): 52–67.

MacLeod, Roy, ed. *Government and Expertise: Specialists, Administrators and Professionals, 1860–1919.* Cambridge: Cambridge University Press, 2003.

MacLeod, Roy M. "The Alkali Acts Administration, 1863–84: The Emergence of the Civil Scientist." *Victorian Studies* 9, no. 2 (1965): 85–112.

——. "Government and Resource Conservation: The Salmon Acts Administration, 1860–1886." *Journal of British Studies* 7, no. 2 (May 1, 1968): 114–50.

MacLeod, Roy M., and P. D. B. Collins. *The Parliament of Science: The British Association for the Advacement of Science 1831–1981.* Northwood: Science Reviews, 1981.

MacNalty, Arthur S. "Emil von Behring." *British Medical Journal* 1, no. 4863 (March 20, 1954): 668–70.

MacQuitty, Betty. *The Battle for Oblivion: The Discovery of Anaesthesia.* London: Harrap, 1969.

Manuel, Diana E. *Marshall Hall (1790–1857): Science and Medicine in Early Victorian Society/Diana E. Manuel.* Atlanta: Rodopi, 1996.

Mayer, Jed. "The Expression of the Emotions in Man and Laboratory Animals." *Victorian Studies* 50, no. 3 (2008): 399–417.

"Medico-Legal." *British Medical Journal* 1, no. 2730 (1913): 917–18.

Moscoso, Javier. *Pain: A Cultural History*. Houndmills: Palgrave Macmillan, 2012.

Mott, F. W., and C. S. Sherrington. "Experiments upon the Influence of Sensory Nerves upon Movement and Nutrition of the Limbs. Preliminary Communication." *Proceedings of the Royal Society of London* 57 (January 1, 1894): 481–88.

Murray, George R. "Some Effects of Thyroidectomy in Lower Animals." *British Medical Journal* 1, no. 1830 (January 25, 1896): 204–6.

Nussbaum, Martha C. *Frontiers of Justice: Disability, Nationality, Species Membership*. Cambridge, MA: Belknap Press of Harvard University Press, 2007.

——. "The Moral Status of Animals." *Chronicle of Higher Education* 52, no. 22 (February 3, 2006): B6–B8.

Otis, Laura. "Howled out the of Country: Wilkie Collins and H.G. Wells Retry David Ferrier." In *Neurology and Literature, 1860–1920*, edited by Anne Stiles, 27–51. Houndmills: Palgrave Macmillan, 2007.

Otter, Chris. "Civilizing Slaughter: The Development of the British Public Abattoir, 1850–1910." In *Meat, Modernity, and the Rise of the Slaughterhouse*, edited by Paula Young Lee, 89–106. Durham: University of New Hampshire Press, 2008.

Ozer, Mark N. "The British Vivisection Controversy." *Bulletin of the History of Medicine* 40, no. 2 (1966): 158–67.

Parris, Henry. "The Nineteenth-Century Revolution in Government: A Reappraisal Reappraised." *Historical Journal* 3, no. 1 (1960): 17–37.

Pawley, Emily. "The Point of Perfection: Cattle Portraiture, Bloodlines, and the Meaning of Breeding, 1760–1860." *Journal of the Early Republic* 36, no. 1 (February 25, 2016): 37–72.

Pellew, Jill. "Law and Order: Expertise and the Victorian Home Office." In *Government and Expertise: Specialists, Administrators and Professionals, 1860–1919*, edited by Roy MacLeod, 59–72. Cambridge: Cambridge University Press, 2003.

——. *The Home Office, 1848–1914, from Clerks to Bureaucrats*. Rutherford, NJ: Fairleigh Dickinson University Press, 1982.

Pellew, Jill H. "The Home Office and the Explosives Act of 1875." *Victorian Studies* 18, no. 2 (December 1, 1974): 175–94.

Pemberton, Neil, and Michael Worboys. *Mad Dogs and Englishmen: Rabies in Britain, 1830–2000*. Basingstoke: Palgrave Macmillan, 2007.

Pernick, Martin S. *A Calculus of Suffering: Pain, Professionalism and Anesthesia in Nineteenth-Century America*. New York: Columbia University Press, 1987.

Petryna, Adriana. *Life Exposed: Biological Citizens after Chernobyl*. Princeton, NJ: Princeton University Press, 2013.

Philanthropos. *Physiological Cruelty, Or, Fact V. Fancy: An Inquiry Into the Vivisection Question*. London: John Wiley and Sons, 1883.

Pick, Anat. *Creaturely Poetics: Animality and Vulnerability in Literature and Film*. New York: Columbia University Press, 2011.

Porter, Roy. *Flesh in the Age of Reason: The Modern Foundations of Body and Soul*. New York: W. W. Norton, 2005.

Porter, Theodore M. *Trust in Numbers*. Princeton: Princeton University Press, 1996.

Rader, Karen. *Making Mice: Standardizing Animals for American Biomedical Research, 1900–1955*. Princeton: Princeton University Press, 2004.

Radford, Mike. *Animal Welfare Law in Britain: Regulation and Responsibility*. Oxford: Oxford University Press, 2001.

"Report of the Commission on Experimentation on Animals." *British Medical Journal* 1, no. 790 (February 19, 1876): 227–28.

"Report on Swine-Fever in Great Britain." London: Printed by Eyre and Spottiswoode, 1886.

"Report to the Secretary of State on the Framework of Legislation to Replace the Cruelty to Animals Act 1876." London: Home Office, 1981.

Regan, Tom. *The Case for Animal Rights*. Berkeley: University of California Press, 1983.

Rey, Roselyne. *The History of Pain*. Cambridge, MA: Harvard University Press, 1998.

Rhodes, Gerald. *Inspectorates in British Government: Law Enforcement and Standards of Efficiency*. London; Boston: Allen & Unwin for the Royal Institute of Public Administration, 1981.

Richards, Stewart. "Anaesthetics, Ethics and Aesthetics: Vivisection in the Late Nineteenth-Century British Laboratory." In *The Laboratory Revolution in Medicine*, edited by Andrew Cunningham and Perry Williams, 142–69. Cambridge; Cambridge University Press, 2002.

———. "Drawing the Life-Blood of Physiology: Vivisection and the Physiologists' Dilemma, 1870–1900." *Annals of Science* 43, no. 1 (January 1986): 27–56.

Richardson, Ruth. *Death, Dissection and the Destitute*. Chicago: University of Chicago Press, 2001.

Ritvo, Harriet. *The Animal Estate: The English and Other Creatures in the Victorian Age*. Cambridge, MA: Harvard University Press, 1987.

———. *The Platypus and the Mermaid: And Other Figments of the Classifying Imagination*. Harvard University Press, 1998.

———. "Possessing Mother Nature: Genetic Capital in 18th-Century Britain." In *Early Modern Conceptions of Property*, edited by John Brewer and Susan Staves, 413–26. London: Routledge, 2014.

———. "Pride and Pedigree: The Evolution of the Victorian Dog Fancy." *Victorian Studies* 29, no. 2 (January 1, 1986): 227–53.

Roberts, David. "Jeremy Bentham and the Victorian Administrative State." *Victorian Studies* 2, no. 3 (1959): 193–210.

Robinson, Victor. *Victory over Pain, a History of Anesthesia . . .* New York: Schuman 1946.

Royal Commission on Vivisection. "Report of the Royal Commission on the Practice of Subjecting Live Animals to Experiments for Scientific Purposes." London: Printed by G.E. Eyre & W. Spottiswoode, for H.M. Stationery off., 1876.

Royal Commission on Vivisection. "Appendix to First Report of the Commissioners : Minutes of Evidence, October to December, 1906." London: printed for H.M.S.O. by Wyman & Sons, 1907.

———. "Final Report of the Royal Commission on Vivisection." London: Printed by Wyman & Sons for H.M. Stationery off., 1912.

———. "Appendix to Fourth Report of the Commissioners : Minutes of Evidence, October to December, 1907." London: printed for H.M.S.O. by Wyman & Sons, 1907.

Sanger, Carol. "Legistlating with Affect: Emotions and Legistlative Law Making." In *Passions and Emotions: NOMOS LIII*, edited by James E. Fleming, 38–76. New York: New York University Press, 2012.

Scarry, Elaine. *The Body in Pain: The Making and Unmaking of the World*. New York: Oxford University Press, 1987.

Schmidgen, Henning. "Pictures, Preparations, and Living Processes: The Production of Immediate Visual Perception (Anschauung) in Late-19th-Century Physiology." *Journal of the History of Biology* 37, no. 3 (2004): 477–513.

Shapin, Steven. *The Scientific Life: A Moral History of a Late Modern Vocation*. Chicago: University of Chicago Press, 2010.

Shapin, Steven, and Simon Schaffer. *Leviathan and the Air-Pump: Hobbes, Boyle, and the Experimental Life: Including a Translation of Thomas Hobbes, Dialogus Physicus de Natura Aeris by Simon Schaffer*. Berkeley: University Presses of California, 1985.

Shapiro, Barbara J. "'Fact' and the Proof of Fact in Anglo-American Law (c.1500–1850)." In *How Law Knows*, edited by Austin Sarat, Lawrence Douglas, and Martha Merrill Umphrey, 25–71. Stanford, CA: Stanford University Press, 2007.

Sharpey-Schafer, Edward. "History of the Physiological Society during Its First Fifty Years, 1876–1926." *Journal of Physiology* 64, no. 3, Suppl. (1927): 1–76.

Sherrington, Charles S. "Experiments in Examination of the Peripheral Distribution of the Fibres of the Posterior Roots of Some Spinal Nerves." *Philosophical Transactions of the Royal Society of London. B* 184 (January 1, 1893): 641–763.

Shevelow, Kathryn. *For the Love of Animals: The Rise of the Animal Protection Movement*. New York: Holt Paperbacks, 2009.

Shmuely, Shira. "Curare: The Poisoned Arrow That Entered the Laboratory and Sparked a Moral Debate." *Social History of Medicine* 33, no. 3 (August 1, 2020): 881–97.

——. "Law and the Laboratory: The British Vivisection Inspectorate in the 1890s." *Law & Social Inquiry* 46, no. 4 (November 2021): 933–63.

Shukin, Nicole. *Animal Capital: Rendering Life in Biopolitical Times*. Minneapolis: University of Minnesota Press, 2009.

Silbey, Susan, and Patricia Ewick. "The Architecture of Authority: The Place of Law in the Space of Science." In *The Place of Law*, edited by Austin Sarat, Lawrence Douglas, and Martha Merrill Umphrey, 75–108. Ann Arbor: University of Michigan Press, 2003.

Silbey, Susan S., and Egon Bittner. "The Availability of Law." *Law & Policy* 4, no. 4 (October 1, 1982): 399–434.

Singer, Peter. *Animal Liberation: The Definitive Classic of the Animal Movement*. Reissue. New York: Harper Perennial Modern Classics, 2009.

——. "Reflections." In *The Lives of Animals*, by J. M. Coetzee, 85–92. Princeton: Princeton University Press, 2001.

Smith, Roger. "The Embodiment of Value: C. S. Sherrington and the Cultivation of Science." *British Journal for the History of Science* 33, no. 3 (September 1, 2000): 283–311.

Snow, Stephanie J. *Blessed Days of Anaesthesia: How Anaesthetics Changed the World*. Oxford: Oxford University Press, 2009.

——. *Operations Without Pain: The Practice and Science of Anaesthesia in Victorian Britain*. Basingstoke: Palgrave Macmillan, 2005.

Stiles, Anne. *Popular Fiction and Brain Science in the Late Nineteenth Century*. Cambridge: Cambridge University Press, 2011.

Straley, Jessica. "Love and Vivisection: Wilkie Collins's Experiment in Heart and Science." *Nineteenth-Century Literature* 65, no. 3 (2010): 348–73.

Stilt, Kristen. "Law." In *Critical Terms for Animal Studies*, edited by Lori Gruen, 197–291. Chicago: University of Chicago Press, 2018.

Sturdy, Steve. "Knowing Cases: Biomedicine in Edinburgh, 1887–1920." *Social Studies of Science* 37, no. 5 (October 1, 2007): 659–89.

Swazey, Judith P. "Action Propre and Action Commune: The Localization of Cerebral Function." *Journal of the History of Biology* 3, no. 2 (1970): 213–34.

Sykes, Keith, and John Bunker. *Anaesthesia and the Practice of Medicine: Historical Perspectives*. Boca Raton, FL: CRC Press, 2007.

Tansey, E.M. "'The Queen Has Been Dreadfully Shocked': Aspects of Teaching Experimental Physiology Using . . ." *American Journal of Physiology* 274, no. 6 (June 2, 1998): 18–33.

———. "The Wellcome Physiological Research Laboratories 1894–1904: The Home Office, Pharmaceutical Firms, and Animal Experiments." *Medical History* 33, no. 1 (January 1989): 1–41.

Thomas, Keith. *Man and the Natural World: Changing Attitudes in England, 1500–1800*. New York: Pantheon Books, 1983.

Tigertt, W. D., and William Smith Greenfield. "Anthrax. William Smith Greenfield, M.D., F.R.C.P., Professor Superintendent, the Brown Animal Sanatory Institution (1878–81). Concerning the Priority Due to Him for the Production of the First Vaccine against Anthrax." *Journal of Hygiene* 85, no. 3 (December 1, 1980): 415–20.

Turner, James. *Reckoning with the Beast: Animals, Pain, and Humanity in the Victorian Mind*. Baltimore: Johns Hopkins University Press, 1980.

Victoria Street Society for the Protection of Animals from Vivisection. *Anæsthetics and Vivisection*. London: 1887.

Wailoo, Keith. *Pain: A Political History*. Baltimore: Johns Hopkins University Press, 2014.

Wang, Jessica. "Dogs and the Making of the American State: Voluntary Association, State Power, and the Politics of Animal Control in New York City, 1850–1920." *Journal of American History* 98, no. 4 (March 1, 2012): 998–1024.

———. "Imagining the Administrative State: Legal Pragmatism, Securities Regulation, and New Deal Liberalism." *Journal of Policy History* 17, no. 3 (2005): 257–93.

———. *Mad Dogs and Other New Yorkers: Rabies, Medicine, and Society in an American Metropolis, 1840–1920*. Illustrated edition. Baltimore: Johns Hopkins University Press, 2019.

Weber, Max. *Economy and Society*. Edited by Guenther Roth and Claus Wittich. Berkeley: University of California Press, 2013.

West, Robin. "The Anti-Empathic Turn." In *Passions and Emotions*, edited by James E. Fleming, 243–88. New York: New York University Press, 2012.

Westacott, Evalyn. *A Century of Vivisection and Anti-Vivisection: A Study of Their Effect upon Science, Medicine and Human Life during the Past Hundred Years*. Ashingdon, Essex: C.W. Daniel Co., 1949.

———. "A New Leader—Miss Emilie Augusta Louise Lind-Af-Hageby." In *A Century of Vivisection and Anti-Vivisection: A Study of Their Effect upon Science, Medicine and Human Life during the Past Hundred Years*, 189–96. Ashingdon: C.W. Daniel Co., 1949.

White, Paul. "Introduction." *Isis* 100, no. 4 (December 1, 2009): 792–97.

——. "Sympathy under the Knife: Experimentation and Emotion in Late Victorain Medicine." In *Medicine, Emotion and Disease, 1700–1950*, edited by Fay Bound Alberti, 100–124. New York: Palgrave Macmillan, 2006.

White, Paul S. "The Experimental Animal in Victorian Britain." In *Thinking with Animals: New Perspectives on Anthropomorphism*, edited by Lorraine Daston and Gregg Mitman, 59–82. New York: Columbia University Press, 2005.

Wilkinson, Lise. *Animals and Disease: An Introduction to the History of Comparative Medicine*. Cambridge: Cambridge University Press, 2005.

Willrich, Michael. *Pox: An American History*. New York: Penguin Press, 2011.

Wolfe, Charles T. "Vitalism and the Resistance to Experimentation on Life in the Eighteenth Century." *Journal of the History of Biology* 46, no. 2 (May 1, 2013): 255–82.

Wolfe, Cary. *What Is Posthumanism?* Minneapolis: University of Minnesota Press, 2009.

Woods, Abigail. *A Manufactured Plague: The History of Foot-and-Mouth Disease in Britain*. London: Routledge, 2013.

Woods, Rebecca J. H. *The Herds Shot Round the World: Native Breeds and the British Empire, 1800–1900*. Chapel Hill: University of North Carolina Press, 2017.

Yeo, Gerald F. "Note on the Application of the Antiseptic Method of Dressing to Craniocerebral Surgery." *British Medical Journal* 1, no. 1063 (1881): 763–64.

INDEX

Ingram Content Group UK Ltd.
Milton Keynes UK
UKHW012023080623
423123UK00005B/305